ALSO BY THE AUTHOR

Last Stand at Rosebud Creek: The Story of Eighteen People & a Power Plant

The Boys
BEHIND THE BOMBS

The Boys
BEHIND THE BOMBS

Michael Parfit

Little, Brown and Company

Boston   Toronto

FIRST EDITION

Library of Congress Cataloging in Publication Data

Parfit, Michael.
    The boys behind the bombs.

    1. Strategic forces — United States. 2. Atomic weapons. I. Title.
UA23.P277 1983   358'.17.'0973   83-786
ISBN 0-316-69057-0

BP

Designed by Dede Cummings

*Published simultaneously in Canada
by Little, Brown & Company (Canada) Limited*

PRINTED IN THE UNITED STATES OF AMERICA

To Debbie

CONTENTS

ONE

Penetration

1

ALBERT LATTER had heard of Los Alamos. It was a barren, empty place in high desert. He was twenty-one. He was a tall, slender young man with dark eyes. He was becoming a physicist at the University of California and wanted to join the Navy. After Pearl Harbor he had, in fact, applied to enlist and go to radar school. Somehow the Navy hadn't called back. It didn't take him too long to figure out that he had been impressed into Dr. Robert Oppenheimer's unusual project. He would not have recognized the code name the project later acquired: Manhattan. But now everyone was going to Los Alamos. He didn't want to go to Los Alamos. Winter wind, the blast of summer. No girls. No girls! But Dr. E. O. Lawrence was staying in Berkeley. Latter was acquainted with Dr. Lawrence. He had a few quiet words with him. And when the trains left for New Mexico, for that parched life on the edge of discovery, Albert Latter remained behind.

So he missed it, the greatest invention of humankind.

2

I

DR. ALBERT LATTER strode across the small room and searched the hall. No one was there. He walked back to the south window and searched the Pacific Ocean. Its emptiness shone. It was late in the morning. Latter carried a long wooden pointer. Its tip was a white rubber cone that looked like a model of the warhead of an intercontinental ballistic missile. Behind him on a blackboard were hasty sketches: lines, circles, a graph with diverging lines. In front of him sat a journalist who was taking notes by hand because the use of a tape recorder would have set off electronic alarms. The journalist's name was Michael. Latter flexed the pointer between his hands, put it down, and picked up a piece of chalk. Latter was as slender as the pointer. His fingers were as cool and thin as the chalk. He walked to the west window and searched the Hollywood Hills.

"We have spent a lot of money on national defense," he said. "But of one hundred eighty billion dollars a year we spend a small fraction on strategic war." He walked across the room. His eyes were small and dark.

Latter protected them, squinting in the windows' glare. He sat down. He spoke again:

"It's hard to imagine fighting the Armageddon war, so we don't want to spend money on it. And what money we do spend is often spent more on appearances instead of on actual war fighting capability."

Michael leaned forward. He stared at the physicist. His own eyes were as large as some creature of the night. They were open wide, as if where Latter saw light Michael saw darkness. He glanced down at the papers before him. One of the sheets of paper bore the physicist's credentials, a vague and dry life history of achievement, a career entwined with the growth and development of nuclear weapons. He looked up again at the physicist, his face blank.

Latter walked to the blackboard and put the chalk down in the tray. He picked up the pointer. He sat down in a chair.

"If you want a weapons system, you'd better believe it is in response to something," Latter said. "Unless you recognize the threat you will never find a solution. It is very hard to get threats recognized in the United States."

He stood up and walked to the door. He looked out into the hallway.

"In nineteen sixty-three," he said, "we invented the threat."

Michael scribbled. Finished, he looked up, waiting. Latter's eyes seemed hungry. Again they swept across the Pacific. He stood up, walked across the room, and sat down in an opposite chair. He had the face of a raptor — a hawk, a falcon. His nose was large, slightly bent. His white hair was swept back from his face as if he flew.

Latter's eyes were narrowed. Was the whole room too

bright, or had the sight of the detonation of a four-megaton bomb fifty miles high over Johnston Island back in 1958 permanently dazzled him? He got up and walked across the room. He searched the sky, the mountains and the sea. Michael watched him, saying nothing, his notebook momentarily abandoned, watching the man's face. What was he searching for, with such restless eyes, with such hunger? For a moment Michael, too, followed Latter's gaze and stared out to sea.

II

"Gentlemen," the Strategic Air Command intelligence briefer said, "during the past decade the Soviet Union has been involved in a relentless and unabated buildup of its armed forces." The briefer was dressed in an immaculate Air Force dress uniform. He wore a small bright patch of ribbons, and the silver bar of a first lieutenant. His dark hair was short. His eyes were dark and steady. Michael watched him, making notes in a steno pad. It was late in the afternoon. On the screen on the wall appeared several faces, outlined in red.

"This Soviet High Command has directed these modernization programs," the briefer said. He pointed at each face with a light pointer. It made it seem that a small, bright fly was leaping across the screen.

"Under the Soviet president is D. F. Ustinov, minister of defense. Next is V. G. Kulikov, commander in chief of the Warsaw Pact." Kulikov stared out of the slide, up and to his left, his right eyebrow slightly raised. "On his right is N. V. Ogarkov, chief of the general staff." Ogarkov had an abundance of wavy hair. He, too, stared up and to the left. The bright fly jumped. "On his right, S. L. Sokolov, first deputy, ministry of defense." Sokolov, all photographic grain, looked straight into the camera.

The slide vanished. A chart appeared.

"Analysis of Soviet defense expenditures has shown an average real growth of about three to five percent per year," the briefer said. His voice was emotionless. He looked about twenty-seven years old. He had given the briefing often. "At one hundred seventy-five billion dollars, the estimated cost of Soviet defense activities in nineteen eighty was fifty percent higher than the U.S. outlay of one hundred fifteen billion dollars."

Slides changed silently. On the screen appeared a collage of black and white photographs. Michael, sitting in the back of the briefing room, made notes. There was a swept wing bomber, two missiles firing, a submarine, and a Soviet crest: a star, a globe, a hammer and sickle. They all had the vague, smudged look of pictures taken in stealth.

"Like the United States," the briefer continued, "the Soviet Union's strategic nuclear posture features a triad of forces: land-based intercontinental ballistic missiles, ballistic missile submarines and their missiles, and bombers." The screen now showed a graph.

"The U.S.S.R. currently maintains a numerical superiority over the U.S. in ICBMs and submarine launched missiles. Since 1966 the Soviet Union has developed six new types of ICBM in comparison to one new ICBM for the United States. Research and development in all military-related fields shows continued disparity."

The screen changed. The briefer bounced slightly in his bright shoes. His speech was hard and quick. Michael wrote in his book. "Continued disparity," then stopped abruptly, looking up at the shiny man.

"Over the past decade," the briefer continued briskly, "Soviet expenditures for ICBMs were about four times those of the U.S. The privileged position of ICBMs in the U.S.S.R. is supported by an elaborate and influential military-industrial complex. Our intelligence collection systems indicate that four design bureaus develop Soviet ICBMs. Missile production takes place at several main assembly plants and hundreds of subassembly plants, employing thousands of workers."

The screen changed. The slide was headed "Strategic Force Composition." It contained six small pie graphs, but it was dominated by two faces in profile: on the left a profile cut from the red and white colors of the Soviet flag; on the right a profile shaped of red and white stripes with a rectangle of white stars on blue. The two faces were identical: high foreheads, heavy brows, and massive, determined chins. They faced each other, nose to nose.

"The Soviets are believed to be testing a new and improved Anti-Ballistic Missile system," the briefer said. "They are working on a new anti-satellite program. They are developing a new space booster capable of launching very heavy payloads into orbit, including laser weapons. With their new SS-18 and SS-19 intercontinental ballistic missiles the Soviet Union is technically capable of threatening an incapacitating attack on United States Minuteman Missiles. For this reason the Air Force has recommended the immediate deployment of the MX missile."

"Gentlemen," the briefer said mechanically, "the United States has no recourse but to respond."

Michael wrote the words across his page. The briefer, in his pause, studied Michael. Michael's shoes were spit-shined, too. His hair was short. He wore blue slacks and a blue shirt and tie. He stared back at the briefer. Then both looked back at the screen, at the erect silhouette of an intercontinental missile.

3

INTERCONTINENTAL BALLISTIC MISSILES

| *United States:* | *1054* |
| *Soviet Union:* | *1398* |

A GENTLE rain fell on California. The mountains drank. Michael drove on a wet highway between hills that were bright with new grass, poppies, and mustard. Reborn streams ran brown under bridges. The air smelled of earth and sea. Pale gray clouds embraced the ridges, and left their moisture on the shiny leaves of the oak forests. Michael passed through the town of Lompoc, where men farm flowers, and came to Vandenberg Air Force Base, on the coast.

It was a place of low brush, sand, mud, and eucalyptus trees, which dripped. Michael signed forms, received a pass, and was taken by a public relations captain to a large briefing room, where he and the captain sat alone. In moments they were joined by a stocky, cheerful major, a representative of the 6595th Missile Test group. The 6595th would eventually be in charge of testing the MX missile. The major vigorously embarked on a briefing. Michael watched. When the major paused, Michael wrote in his notebook:

"Follow-on test mode. Only implement ten to 20 flights. DT & E — with some OT & E."

The major talked for several more minutes. Michael nodded at appropriate occasions. He wrote in his book: "Milestone III, around June, 1983. GEMS. ENEC. Nozzle pops out. Cold Gas Ejection."

The major developed a brisk enthusiasm. Michael watched a series of slides of charts, graphs, and tables speed past on a screen. He nodded more. He wrote:

"Hot Flyout. More throw weight. Ten Mark 12 As, or 10 ABRVs. Gems maneuver. Play around in the sky until right position. STML. Mass simulator. Two mass simulators."

The major showed a series of slides of unfinished buildings: enormous cranes, steel beams, scaffolding, all silhouetted against clear California skies. Michael wrote busily:

"Cluster maintenance facilities. ITF. PAB. MMF. GTM. MAB is 140 feet above the ground and 85 feet underground. One thousand people working on DT & E."

The major paused. He looked down at Michael. Michael covered the gibberish on his pad with his arm.

"Any questions so far?" the major asked.

Michael glanced at the book, then looked the major straight in the eye.

"No," he said.

The major smiled. Michael smiled back. The major said, quite clearly:

"In testing a missile you have to look at what you are attempting to do. It is a misnomer to talk about throw weight, because it means nothing without accuracy." He grinned.

"I can shoot you in the shoulder with a three fifty-seven magnum and stop you," he said cheerfully. Michael looked up at him, eyes huge. The major went on, his smile entirely benign. "I can also shoot you with a twenty-two and stop you. It all depends on accuracy."

Michael wrote it down.

In the rain, with the captain, Michael drove through the low hills of the Air Force base. Between patches of bush lupine — not yet in flower — and iceplant, the soil was a damp, sandy brown. The captain confessed a desire to write novels. Michael expressed sympathy. They came to a high spot in the road. There, across the damp scrub was a yellow framework of steel: two rafters, two sides. It looked like a child's sketch of a house done in yellow crayon on gray.

"That's the MAB," the captain said. "The Missile Assembly Building."

He stopped the car. There was no other traffic on the road. Wind off the Pacific rushed around them. In the distance a grove of eucalyptus trees rippled and heaved before it. Rain glittered on the windshield. Two cranes attended the MAB, adding sticks to the framework. Morning glory flowered in the brush. Off on the side of the MAB framework Michael could see a tiny spark: a welder, working in the rain.

The captain took Michael to lunch in the officer's club. In the large, noisy room there was a mockup of the fourth stage of the MX missile. The fourth stage was the final stage, the nosecone. It was white, 92 inches in diameter at the base and about 15 feet high. In the dining room it seemed enormous. It looked like the prow of a ship, cleaving the air above both the hungry and the well-fed. It looked like the front end of a gigantic spear. It looked like a grossly enlarged .22 slug. It looked like an enormous canine fang.

No one else even seemed to notice the nosecone. It

was just a part of the furniture. Men in blue uniforms mingled with civilians in its vicinity. No one spent much time looking at it. They walked around it. Michael loaded a tray with a plate full of enchiladas and walked around it, too. He sat at a table with the captain. He watched the head of the missile, carefully, cautiously, as if its shape would reveal some secret of its intent.

"It's a four-stage missile," the captain said. He was a mild, friendly black man. He had spoken again, wistfully, about his novels. None had yet been published. "The MX weighs one hundred ninety thousand pounds," he said. "It'll carry ten RVs. Sorry, Reentry Vehicles — warheads. The Minuteman III, the most up to date missile today, weighs seventy-eight thousand pounds, and carries three RVs. Quite a difference. The RVs are independently targetable. They ride on the fourth stage, which is called the bus. The bus drives through space, and at the right point, it kicks off each RV, and the RV penetrates to the target."

Michael eyed the white cone. The room was full of slender men in blue uniforms, all paying it absolutely no attention. Michael asked:

"What's gas ejection mean?"

"Oh, yeah. It gets blown out of its launcher by steam, instead of rocket propellant. Less damage to the launcher."

The two men were silent for a few minutes, eating. Michael kept his eye on the white nosecone, the bus. An officer passed close beside it and instinctively ran his hand along its gleaming white skin. The captain spoke:

"Oh, yeah. And another thing. The MX has a new guidance system. It's called AIRS. I'm not sure what that stands for. What it means is that the MX is a hell of a lot more accurate."

4

I

"The project officer for the MX is Colonel Bill Crabtree," Neil Buttimer said on the telephone. Lieutenant Colonel Buttimer was the public relations man for the Ballistic Missile office at Norton Air Force Base. Buttimer was a man of muffling mildness. "Colonel Crabtree is extremely bright. He's very precise." Buttimer said with interest: "He is very oriented toward the mission." Michal wrote in his book. "Bill Crabtree."

II

"No fire," Jack Hilden told the guard at the gate of the Thiokol Corporation's manufacturing area. He might have been talking about himself. The Great Salt Lake desert stretched in magnificent desolation towards mountains in all directions. The northwest wind smelled of sagebrush. On the hillside in the distance several rows of small buildings stood, each set well apart from the other. On two were flashing red lights. Within them fuel was mixed or housed: the solid propellant used in the missiles Thiokol built and in the United States Space Shuttle.

They were separated so that if one went up the others wouldn't join in, howling in flaming chorus on the hill.

Thiokol built the MX first stage. Hilden was a large, gray man. He seemed a gentle man, as if he spent all his fire putting heat in the bottles his company built on the hillside.

Hilden took Michael into the room where an MX first stage for testing was being made. It was appallingly quiet. Michael looked around. A radio played a lugubrious song called "Torn between Two Lovers." There was no clangor. It was the quietest manufacturing plant Michael had ever visited. Hilden smiled. "I can show you some machine shops if you want," he said apologetically. Michael shook his head. He was watching the bottle grow.

It looked like a huge golden ball of yarn on a mandril. It turned slowly, silently, pulling gold thread from bobbins which moved back and forth along the length of the bottle to lay the thread in shining crisscross patterns. The thread was called Kevlar. The bottle was twenty feet long. The bobbins worked automatically. They were attended by a man in a blue lab coat who worked steadily but with no urgency at all. He and the spindles just walked gently back and forth, spinning a vast golden bead.

"You know," Hilden said. "There's a kind of continous community on this. People know each other. But it was John Hepfer, out at BMO, General Hepfer — He carried the MX from a name to a weapons system." Michael wrote: "John Hepfer."

III

In his office at TRW Systems, Inc., Dr. Allen Schaffer was surrounded by so much potted foliage that in remembering his visit later Michael could have sworn that Schaffer's abundant growth of hair was green, too. It wasn't. It was gray gone nearly white. It was wavy. It swept almost to his shoulders, in contrast to Schaffer's strongly masculine face.

Michael sat beside a large ficus plant. He peered around the leaves, like a roosting owl. Schaffer was a vice president for TRW, the company that had an interlocking relationship with the Air Force on the MX missile. TRW worked on an open-ended contract to provide what is called systems engineering and technical support to the military men. Schaffer spoke of history with a kind of curious weariness. He seemed surprised to find that important events in which he participated interested him so little. But he spoke of the men with a more cheerful eye.

"Some good, objective work was done," he said. "Bill Crabtree was a bright, objective officer. Hepfer went out and recruited him from Systems Analysis. Crabtree was the quarterback of the MX."

IV

At 10 A.M. all the chairs from Room 1302 in the Longworth Office Building of the U.S. House of Representatives were inexplicably stacked outside the room when Michael and a few dozen others arrived to attend a hearing on the MX missile. The room was bare and smelled of industrial cleaner. It seemed to have just been harshly purified.

For a while the crowd surged around in the hall as if shocked by this unforeseen difficulty. Then, seizing initiative, a few of the more aggressive members of the group grabbed chairs from the stack and carried them into the room themselves. The rest of the crowd followed.

There was a murmur of jokes about BYOC parties. The rows they set up inside were ragged.

Michael sat approximately behind a slender, bearded man who seemed unconcerned by the proceedings. Everyone else in the room wore either an air of slightly worried determination or, in the case of some of the congressmen, of jocund irritation. The central player in the hearing was a man named Peter J. Sharfman, a representative of the Congressional Office of Technology Assessment. Sharfman was a placid but sharply articulate man who delivered a lengthy analysis of the current proposals for basing the MX missile and spoke of none with enthusiasm.

"We were requested to assume that the MX is required for our national defense," Sharfman said. He wore a faint air of controlling skepticism. He proceeded to outline options: Put them in submarines; put them in airplanes; put them in vertical silos; defend them with unproven anti-missile missiles; put them on racetracks, in multiple protective shelters.

Michael glanced over the shoulder of the man in front of him. His face was striking. It was uncommonly lean. It wore a ragged beard, and a kind of windblown ruddiness. It was the face of a buccaneer. The man had brought a notebook, but made few notes, writing just once or twice in an upper corner in script so small Michael could make no sense of it.

The hearing was calm. Once in a while a Nevada Congressman, Jim Santini, made unpleasant remarks. Santini, a white-haired man with a round face and dark eyebrows that gave him a bright-eyed look, like a bold chipmunk, picked on Sharfman's description of why the Air Force could not be expected to like submarines:

"When you cite institutional restraints, that's delicate language for bias, isn't it? The legislative branch wallows

in institutional restraints. I hope you will jar us loose."
Sharfman, unjarrable himself, rolled on, causing a stir
in the room only once: The alleged window to Russian
attack, he said, might yawn throughout the '80s:

"We are skeptical that any survivable MX basing can
be built within the current decade," Sharfman said.

The audience reacted to Sharfman's words with a faint
stir, a rustle that was as close as you could get in these
halls to a collective gasp. The man in front of Michael
did not join the restrained consternation. He doodled
minutely on his page.

But when the hearing was over and little clusters of
people stood around the room, weighing down their individual seriousness with group discussion while the congressmen fled, the bearded man laughed. He was talking
with a young woman who appeared to idolize him. It
was a bold laugh. It was a rattling good laugh. It rang
against the high ceilings and the yellow walls. It did not
seem entirely driven by good humor. There was hardness in it.

"Who's he?" Michael asked a visiting Friends of the
Earth lobbyist. Was he one of the more picturesque
opponents of the MX missile, shipped in from some
windseared mine in Nevada?

"That's Marv Atkins," said the lobbyist. "He's from
the Pentagon. In D.D.R. and E."

Michael wrote down the name. "Marvin Atkins, office
of the Director of Defense Research and Engineering."
Dr. Marvin Atkins, he found out later, was an expert in
the effects of nuclear weapons.

V

On the walls of Gary Aubert's office there was a huge
map of the Union of Soviet Socialist Republics. Aubert

was part of Albert Latter's firm, Research and Development Associates, the think tank. He did not have a view of the Pacific. Instead he had the map.

"The map doesn't mean anything," Aubert told Michael casually. In the map were stuck three or four small red pins. Aubert noticed Michael trying to read the names of the pinned cities. They seemed to be rural, but that's as far as Michael got. "The pins don't mean anything, either," Aubert said.

Aubert had a short beard and cold blue eyes. He looked as calm and hard as an ice-hockey forward. He spoke of Bill Crabtree.

"It's amazing the dedication of these guys, with the sort of pressures that are put on them," he said. "Bill Crabtree is one of the most dedicated Air Force guys I know. He probably lives on the red-eye special to Washington to give briefings. He probably has as much responsibility for keeping the thing on track as anybody. He's tenacious, conscientious. He has given body and soul to this project."

VI

In Santa Barbara, California, rain blasted the night. The governor of Utah had come to California to speak in opposition to the MX missile, and Michael drove through the rain to hear him. The parking lot was nearly empty. The dark of the night seemed polished by the rain, which glittered in the streetlights and on the shiny black of the pavement. Michael waited in the car for a pause in the torrent. The water fell steadily. It was as if the long, balanced relationship between earth and sky had finally disintegrated and was collapsing in flood. It was impossible that this torrent could continue. But it did. After five minutes he gave up and ran the fifty feet to the doorway. When he reached shelter, gasping, he was soaked.

Inside, Utah Governor Scott Matheson was already on the stage, warning the small audience not to be too dazzled by the glare of the overhead lights off the polished dome of his forehead, preparing to warn them still more against the dazzling promises of the MX missile. The audience, four dozen high school and college social studies students, giggled softly.

On the California stage, with wild-looking students in the audience and with the world dissolving in flood outside, Matheson looked like the last righteous prophet of the Church of Latter-Day Saints. He was slender and upright. He wore a dark suit, a white shirt, and a blue tie. His limited gray hair and his gray mustache were closely trimmed. He wore the expression of the elder Mormon Stake bishop: benign, stern, thoughtful.

His speech was not flat, but its highs and lows were predictable, with no excesses to reveal genuine heat. He spoke of the formal arguments that sustained or opposed the placement of MX in the racetrack base in Utah and Nevada. The window of vulnerability. Whether or not the Air Force could really hide an enormous, heavy missile that ticked like Peter Pan's crocodile with electronics and radioactivity. Putting the missiles on submarines. The environmental and social impact upon the land and people of Utah. Near the end of the otherwise unremarkable speech he paused and looked wearily out into the tiny group.

"It appears evident that most significant scientific and political forces are committed that we need to build the missile," he said, leaning on a podium purchased by the class of 1962. "The reality of the missile seems to be a given."

The students appeared satisfied. Acronyms in glorious

quantities had tumbled from the stage to be scooped up by those who needed to pin them, like medals, to the front of their discourse and so claim honor and authority. But Matheson had not revealed why he should spend his effort on such an insignificant engagement.

The rain still fell. The cars should all have floated away. Michael drifted towards the doors with the little cluster of people that surrounded Matheson. Matheson nodded and shook hands with people who seemed warmed in reflection. He, too, glanced outside. He was due to fly off in a small aircraft that evening. It must have seemed a nightmare to him to imagine soon becoming engulfed in that collapsing night.

Michael worked himself through the little crowd and added his handshake and name to the meaningless list. He had one question. Matheson agreed to answer, but glanced toward the door. His limousine had just arrived at the flooded curb. He turned back with a concealed twitch of impatience. Lord, to get this night over!

"Who do you think — what in your estimation —" Michael said. Matheson began a back-away shuffle. Michael rushed on. "Who — Who is the champion of MX?"

Matheson looked at him sharply. There was a hard glitter in his eyes. Quickly, clearly, before he turned to leave, he said:

"Seymour Zeiberg."

Outside the door, on the edge of the slash of rain, Michael wrote in his book. "Seymour Zeiberg."

VII

In a retired general's office in Cambridge, Mass., Bill Crabtree was the first item of conversation.

"Bill Crabtree has had this thing on his back for five years," said the general. "He has worked, I think, more than any human being should be asked to work." The general was a reflective and colloquial man. He spoke in between mouthfuls of pipe smoke. Missiles and space shuttles soared from his walls in photographs of fire and steam. He had no haste in his words, but no shortage of them, either. What, Michael asked, was the reason for that kind of — why did someone do that?

"Why?" asked the general. "Why does anyone want to be a soldier?" He looked briefly at Michael, not expecting a reply. Michael's eyes were bland. "No question about it," the general went on. "You couldn't possibly take the punishment you take in those jobs if you didn't feel somethin' much deeper. You certainly don't get paid enough to make up for the kind of effort you have to put in to do those jobs. Then you're frequently in that never-never land between the dedication that it takes personally to put that much effort into a job and the public's reaction to the military between wars, which is not very . . ." The general let a sad shake of the head finish the sentence. "Bill Crabtree," he said, "is a very, very competent, very dedicated, very loyal person."

VIII

The famous physicist smiled as Michael asked him a question. He replied:

"So far as the United States is concerned, how does a person develop confidence that we're doing something sensible? It has to be based on that person's judgment of the people who are in control."

Michael read him a list of names. Bill Crabtree. Marvin Atkins. John Hepfer. Albert Latter. The physicist smiled. He said:

"There's another person that you maybe ought to focus on. He is much more of an overall kind of person. He's quite outspoken, and expresses his views very freely. He's very, ah —"

"Forthright?"

"Very — Yes. He's ready to express himself, and expresses himself quite strongly. General Jasper Welch."

IX

In a small cafeteria over shops on the bright side of the street somewhere in Santa Monica, California, Michael met by appointment with a man whom Michael thought of as the Rand Pinwheel. The pinwheel itself was the little circular slide rule invented by the Rand Corporation that described the effects of nuclear explosions. If you were two miles from the explosion of an MX warhead, and you had a Rand Pinwheel, it could tell you that the wind would soon reach about 250 miles an hour and that any exposed flesh would receive third-degree burns. The man Michael met had once worked for Rand, and was a small source of information, most of it depressing.

The two men had sandwiches and salads. The Rand Pinwheel demanded anonymity. Michael agreed, reluctantly. Pinwheel was a man of tidy bulk, slightly graying. He and Michael both had a glass of red wine. Pinwheel was enthusiastically cynical.

"You remember *The Naked City* TV series?" he asked. "At the end there was a wide-angle view of New York City and the narrator would say, 'There are seven million stories out there.' Well, there are at least that many stories about the MX missile and the truth lies nowhere among them." Pinwheel didn't mind Michael writing down his remarks as long as he would not be associated with them. He rejected a tape recorder, perhaps assuming that he could always claim to be misquoted but that tapes were hard to deny.

He boomed along, engaging himself in the marvelous debate which so preoccupied almost everyone Michael met: Where you should put the MX missile. Basing modes. Land basing, sea basing, and airmobile basing. He laughed about the plan to scatter the missiles across Nevada and Utah. "Oh, it's part of the physicist's absolute belief that there is always a gimmick," he said with scorn. Pinwheel was a political scientist. "That's the physicist's dream. It's another one of those crazy options that came out of RDA. Al Latter and the boys."

X

Michael called Neil Buttimer again.

"I'd like to get an interview with Bill Crabtree," he said. "I'll be in the area for a couple of weeks."

"I'm afraid Colonel Crabtree's on TDY for most of that period," Buttimer said pleasantly. "But I'll see what I can do."

XI

The lights of Salt Lake City glowed up on scattered clouds. There was clear air in the basin, the fresh, bright air of winter. Cars from California with ski racks were parked at the motels. At the downtown convention center, not far from the shining Mormon temple, the Air

Force held a hearing to gather comments on its draft environmental impact statement for putting the MX missile in Nevada and Utah. Michael was late.

Outside the hearing room a man was standing at a table with a helium bottle blowing up balloons as fast as the machine would let him. WHOOSH. WHOOSH. WHOOSH. They were bright yellow balloons that had "Stop MX" stamped on their sides. They bobbed cheerfully in the filtered air of the convention center as people came by and carried them off into the hearing room. The people wore red buttons, which said "No MX!" The balloons and the buttons made the occasion look like a county fair. It lacked only the limp hot dogs and the acres of Nubian goats.

His eyes were huge in the dimly lit halls. He strode toward the room. He turned a corner and there was Lieutenant Colonel Neil Buttimer. Buttimer gave him a mildly interested smile.

Michael slipped into the huge room alone. He submerged in a sea of yellow balloons. The room was full of people, and full of balloons. They floated over the crowd like cartoon clouds. Tied to hands, they bobbed during applause. One, cast loose, hunted around among the vents on the ceiling like a puppy. Sometimes when a door opened or a small group moved in the room, all the tethered balloons would sway together, in a kind of visible sigh. But the cheerfulness had vanished, and it seemed suddenly that the balloons presided over an immense anger.

One by one speakers advanced to the microphones near the front of the room. The meeting was set up to discuss the specifics of environmental impact; few of the speakers addressed those details. They talked about the military mind. They talked about trust. They talked about nuclear war.

A broad young man with short hair and an immaculate pin-striped suit:

"What I don't want to see is a weapon that sits out in the desert and waits for Armageddon."

A gaunt man in a priest's collar:

"Blessed are the peacemakers, for they shall be called the children of God!"

An older woman:

"May I ask one question: Why can't we and the Soviets talk this out instead of having partial agreements on a lot of things?"

An older man:

"I was in the last war in B29s over Hiroshima and Nagasaki both, right afterward, to take pictures, so I have a really close feeling for that. I haven't stayed with this issue: I got out as quick as I could and forgot the whole thing. I've got some ideas where we could go in another direction, but we don't ever come to reckon with it. We're not willing yet to send our kids out and give some kind of training in trying to work out a way of living with each other. Somewhere we have got to get at it."

Buttimer was still outside, laying a kind of minefield of mild good humor across the entry. Michael, having cleared it, scribbled notes. He never took his eyes from the Air Force men, who sat in front of the room.

They were lined up in a single row, sitting on folding steel chairs, facing the audience. There were ten men. Several were Air Force officers in blue dress uniforms. Two were dark-suited civilian consultants. One was an incredibly long older man in brown slacks, a tan sports jacket, and loafers. He seemed long rather than tall; he was practically supine in the chair, his legs stretched out across an expanse of floor. In his notes, Michael called him "Tan Jacket."

The ten men seemed vulnerable. There was no table in front of them; they could not appear to be working or taking notes or otherwise occupied up there. They sat exposed, facing a vehemence that touched on hate. They looked like targets. Some clearly felt like targets. The two consultants scooted sideways in their chairs, put elbows on knees, crossed their legs and uncrossed their legs. The uniformed officers subconsciously resorted to the power of uniformity, each putting his right ankle on his left knee. Tan Jacket just slumped there in his chair, showing a long, lumpy, white-fringed, big-nosed face to the crowd.

Behind the ten men on a small stage, an Air Force judge presided. After each speaker he stifled any hope of an emotional crescendo in the audience by the placid repetition of the same sentence after each fierce, desperate, satirical, anguished, bitter, or vicious denunciation of the Air Force, the missile, or the government of the United States:

"Thank you; the next card I have before me shows the name of —" The rhythm of his monotone was accompanied by a distant, regular WHOOSH from outside the door. The man with the helium bottle was filling more balloons.

Occasionally one of the uniformed Air Force officers stood to answer a specific question — how much money had been authorized by Congress for the fiscal year; how many acres each missile site would occupy; whether or not the dust stirred up by construction would be a health hazard. But when the speakers talked about war they looked bored. To them all this emotional concern about Armageddon seemed superfluous. It was naive. They wore the exasperation of men for whom these questions had been settled long ago, in rational and unemotional circumstances, by experts. They listened impassively, with concealed contempt. And when called upon to answer a particularly wide-ranging question, they all looked down the row at the man in the tan jacket.

He was the exception in the group. He watched the speakers intently, and appeared to consider every remark. In the general restlessness of the room — the angry crowd, the wandering balloons, the tense officers and the twitching consultants — he was the only wholly relaxed being.

After every speech, Tan Jacket would slowly lift his massive hands and join the crowd's applause. None of the other ten men in the front joined him. At the end of each speech he applauded briefly, and then the judge said: "The next card I have . . ." WHOOSH; another balloon was filled.

Like the consultants, Michael twitched in his chair. He watched Tan Jacket. His applause seemed an odd little gesture in this room full of advocates, a strange token of — courtesy. For a few minutes Michael ceased his endless catchup work of making notes.

A young woman appeared at the microphone. She was young, almost slight, and it looked as if she had been forced against her will to speak by the power of her emotion. She seemed almost in tears.

"I don't understand," she said. "I don't understand. If we have enough missiles to destroy the world, why do we need more? It looks like the answer is that the people who are making the weapons just want to stay in business. That's all. That's why you're doing this to us. You want to make money! I'm more frightened of my own government than I am of the Russians! If you keep doing this it's just a matter of time before there's a big accident. Why are you *doing* this?"

She stopped then, her voice tight. The crowd applauded. Tan Jacket applauded. WHOOSH. The Air Force judge asked the ten men if anyone wanted to respond. "General Toomay?" he said. Tan Jacket stood up.

Toomay stood in front of the crowd, hands in his pockets, slightly stooped to reach down to the microphone. His voice was a bit like his face: large, lumpy, forceful. He spoke with the lazy confidence of a teacher.

"There are many honest, intelligent, good-thinking people who believe there are various ways to cope with the current situation. One group believes we should have unilateral disarmament. I see some of you are here. There are those that believe minimum assured destruction is adequate: they have a number of nuclear weapons, we have a number of weapons; it's a standoff."

It was as if Toomay was giving a half-hour lecture in an intimate room, before friends. "But a third group has been dominating recently." WHOOSH. The friendliness in his voice never hardened. He was just explaining a careful analysis, a logical thought. "They believe that in order to bring the Soviets to the table to negotiate genuine arms reductions, we need to show them we have the will and strength to cope with their monolithic efforts to dominate the world." There was a laugh in the back. There was a WHOOSH from beyond the door. Toomay went on steadily.

"I guess I'm one of those guys. I believe we have to chose between some form of backing down to the Soviets or showing our strength and will."

Toomay sat down, slumping back into the little chair. The crowd rustled; the balloons swayed. Later in the evening while the hearing rambled on Michael saw Toomay standing behind the crowd talking earnestly to the young woman who had expressed such despair. He noticed that he listened, too, and they parted with — courtesy. Michael wrote down the name of the man in the Tan Jacket: Major General John Toomay, retired.

Michael left the hearing before it was over, wandering out with a trickle of the crowd. In the cold, he stood on the sidewalk and watched a single yellow balloon released by a child drift up into the lovely night.

XII

The creek ran clear among white rocks. The spring floods had scoured the channel. There were no weeds in the pools. The emaciated trout lay in the riffles, their gray backs less substantial than their own shadows. Michael walked, crossing the little stream again and again on stones. His feet hurt. He tore his shirt on a twig. He walked on and on, farther and farther away from civilization. There was a river ten miles back in here where he had once found tranquillity.

He was accompanied only by the names in his notebook. Seymour Zeiberg, William Perry, John Toomay, Marvin Atkins, Jasper Welch, William Crabtree, Albert Latter.

He walked on a narrow path. The sun raced across the sky. He sat down beside the path to rest. He looked at the notebook. On the cover, the notebook said, "The Boys behind the Bombs." Inside were the names. It might have been a mistake to bring the book, but the names would have come with him anyway.

Dr. Seymour Zeiberg. Deputy Undersecretary of Defense for Strategic and Space Systems. A native of New York City. The champion of the MX missile. Described as an analytical man. Michael had heard that Zeiberg was a large man. Gary Aubert, with the pins in the map of Russia, had told him Zeiberg was the oldest man he knew who still wore a crewcut. Michael had not met Zeiberg, but he saw him in every seamed and water-stained boulder, a great heavy shape of power. He got up to walk again. He looked up at the sky.

Dr. Marvin Atkins. Director of Offensive and Space Systems for the Deputy Undersecretary for Strategic and Space Systems. Michael saw the high-ceilinged hearing room, and a face with the blood bright in the cheeks and the jawline hidden behind a ragged beard. Steady eyes, and hard laughter. Marvin Atkins, just out of twelve years at the Defense Nuclear Agency, where he had studied the effects of nuclear weapons. Michael kept moving, driving the path under his feet.

Major General John Toomay. Big John Toomay. There were trees out here like Big John Toomay: enormous old snags jutting up out of the streambed. Wind-blown, barkless, incorrigibly rooted to the earth.

Michael climbed the side of a hill. Out of the riparian bottom the hillside was cliff and desert. The cliffs were full of holes. There were the footprints of humans in old mud dried hard after the spring rains; hard as rock almost, the grid of boot treads frozen in the earth, the way they would stay for centuries if they were suddenly filled with ash. Michael walked on. His breath was harsh in his throat, but the emptiness of the land swept the sound away.

Major General John Hepfer. Michael had not yet met John Hepfer. He had been the MX Program Manager for seven years. Who was he — just the engineer, or a driving force? Michael had heard him called "Little John"; in photos most of what he had seen of John Hepfer was an enormous smile.

Major General Jasper Welch. In the photo on his

biography, Welch had slightly drowsy eyes, the kind of eyes you might underestimate. On his chest he wore an unassuming cluster of ribbons, but the biography talked of achievement after achievement. At the age of twenty-four, it said, he "led an experimental nuclear weapon design team which developed the basic design concept still used in most operational systems." Two years later he got his physics Ph.D. He was now Assistant Deputy Chief of Staff, Research, Development and Acquisition, Headquarters, U.S. Air Force, Washington, D.C., which meant he was in the Pentagon.

On the hillside the sun was without mercy. Sweat dripped down Michael's face and into his eyes. He brushed it away. He could taste it on his lips. He walked on. His legs ached. What is it these men share? What devotion, what trait makes them devote their lives to this machine of destruction? What characteristic do they share?

Colonel William Crabtree. MX project engineer. Dedication, dedication, dedication. Michael had not yet met Bill Crabtree. He could see him in nothing — he was an invisible shape, drawing lines and angles and circular error probabilities as the flames leaped. Bill Crabtree.

Dr. Albert Latter. President, Research and Development Associates. The man whose history covers the whole of the life of the bomb. There were hawks in these mountains, wheeling across the sky. Michael could feel that same restlessness in his own legs. His eyes squinted in the sun. He stared through the pine trees on the ridge at the quiet of the valley below. Dr. Albert Latter.

Michael walked five more miles. He came to the river. He had found a calm silence here in other years, ten miles from the parking lot at the end of the road. He took off his pack and sat beside the water. The silence was still there. In the heart of it the water gurgled among the stones.

He stared at the green water, and the water asked him questions. It made noises in the rocks. Will you find the strength to tell the real story that is there? To respect these men who hold in their hands the destruction of the earth? Will you have the grace to allow them to describe their own lives? Will you know how to be just another character among the many? Will there be enough time? Will there be time?

Sy Zeiberg. Big John Toomay. Little John Hepfer. Marv Atkins. Bill Crabtree. Jasper Welch. Al Latter. And who else? The boys behind the bombs. Michael got up and walked aimlessly around the meadow. A breeze stirred the trees. Michael sat down on a rock and looked in his notebook. He drew a tall outline on a page, and wrote "MX" beneath it. He stared at it. It looked like a sketch one might find on the walls of a men's room. He slapped the book shut. He looked at the river. It was green. It slid among the boulders almost in silence. He was planning to stay here two days. He looked at the sun, which was falling. He looked at his watch. He got up abruptly and pulled his pack back on his shoulders. He walked the ten miles back out. By the time he reached the road he was running.

5

INTERCONTINENTAL BALLISTIC MISSILE WARHEADS
United States: *2154*
Soviet Union: *4305*

I

THE spring wind swept the exhaust from the air over the city, and Los Angeles shone. Surfers lay on the water off Malibu like basking seals, waiting for the wind to blow them up some waves. In the mountains the tangles of ceanothus brush, nourished by the rains, sent clouds of blue and white flowers bursting out to dance. At Marina Del Rey the harbor rang with the sound of ropes slatting against aluminum masts, a sound as impatient as the crying of the gulls. Michael, this wind of promise full in his face, walked to the towering building on Admiralty Avenue that rose across the street from the boats, and went up to the tenth floor to see Albert Latter.

He waited in the hall where the secretaries worked. Latter was talking to someone in his office. Through the open door Michael could see him pacing the floor, staring out the window. Michael sat in a chair between two paintings of beached boats. He sat still as a fixture, an inconspicuous man in his blue shirt and tie, his blue slacks, his polished shoes. He folded his hands over his notebook on his lap. His large eyes wandered. The secretaries continued with their work. Beside him was a globe of the world. A bookcase full of telephone books stood against the wall. St. Paul. San Diego. Manhattan. Michael stirred, and asked Phyllis Fisher, Latter's secretary, if the press ever had done any stories on RDA or Latter.

"No," she smiled. "We try to avoid that. We like to keep a low profile."

Latter's talk ended. A gray-haired man left the office, with a quick hard look of curiosity at Michael. Michael returned the glance. Latter came out and shook Michael's hand. Latter's hand was cold. His eyes were alight.

Latter showed Michael into his office. Out the windows spread the town and harbor of Marina Del Rey, on the sunny, seaward side of Los Angeles. It was a view of a glittering blue and white opulence: masts, wrapped sails, shoreline condominiums, elegant small figures on the promenade, who wore crisp white shorts and caps with gold braid, brown skin, gold hair.

Like Latter, Michael stared out the window. The marina gleamed. He looked back into the room. The room was not opulent. Pictures of skiers and black-and-white photographs of beached boats and flowers hung on the wall, along with several modestly framed certificates: "The Department of the Air Force Presents the Decoration for Exceptional Civilian Service." The wall opposite the marina held a blackboard, scuffed with erase marks.

Latter sat down beside Michael, then he rose and looked out the window, his eyes worried in their squint at all that brightness of sunshine and wealth. "This subject," he said, "differs considerably from the history of science." Then he sat down in a chair across the room. "In the scientific world," he went on, "people are careful to remember who said what and what contribution this or that person made to the field." Then he got up again, stood by the door and looked down the hall where the gray-haired man had vanished. "But," he said, "when you ask people the history of the Defense Department recollections are much more subjective. No effort is made to be extremely careful in historical accounts. The subject is surrounded by political factors." He walked to the middle of the room and jammed his hands down into his pockets. He stood there leaning slightly back, looking very thin. "The so-called rational players are often career people whose advancement may depend on the role they played in a multi-million dollar program," he said. "When you ask General So-and-so about something he may have been Colonel So-and-so when he got into the subject. His recollection of his involvement is something you perhaps should not take literally." He sat down. He stood up.

Michael sat in a comfortable chair with his back to the sea. He watched Latter pace the room, past a huge painting on the wall of a skier hunched and flying, near a little, twisted representation of a skier made out of metal on the windowsill. He watched the lean man travel back and forth in the room. Latter was as supple as the metal skier was stiff. It was easy to imagine Albert Latter on the snow. He would be smooth, a sensuous skier, in a cool, dry way, sweeping down the mountain in perfect turns that would make him look slow. He would consume the slope gently, but then he would stand voraciously in the lift line, cold hands gripping the poles, his white hair swept back by the wind, looking around, stamping his skis, poking at the snow; staring up at the hill.

Latter spoke of himself with a kind of deprecating praise. He laughed himself off. He talked about the limited value of "playful people on the outside like me." Then he said: "It was a hobby, inventing these basing systems." He walked over to the window and searched the city. An expression of amused sadness passed across his face.

From 1952 to 1971 Albert Latter worked at the Rand Corporation, in Santa Monica, not far from this tenth-floor office. He had been the head of the Rand physics department. He wrote and organized studies on subjects like missile accuracy, nuclear hardness, and weapons survivability. He invented things — not just mechanical devices, but concepts. For instance, "I said in nineteen fifty-four that X-rays come out of nuclear weapons. That had not been recognized before." Perhaps when he stared in that direction he was searching that ocean of buildings for one familiar facade, to look in the window and see himself turning the knob on the safe of the unknown.

Latter came back and stood beside the blackboard, staring into the hall. He was holding a long piece of white chalk. Without apparent effort, he broke it in half.

"On August twelfth, nineteen sixty-three," he said, "Robert McNamara spoke before the Congress. He spoke on behalf of the Limited Nuclear Test Ban Treaty. It was a dramatic speech."

Michael glanced down at his notebook. Latter sat down in a chair facing Michael. His white hair seemed to stream back from his forehead. He leaned forward.

"McNamara said we didn't need to test in the atmosphere," Latter said. "He said that the people who say that are confused. He said our Minuteman Two missiles could not be effectively attacked because they were dispersed as well as hardened."

Latter got up and walked halfway across the room, jammed his cold hands into his pockets and stared out across the marina. Michael, too, looked out into the bright air, trying to see it: 1963. The year of the signing of the limited nuclear test ban agreement.

Time magazine: "COLD WAR: New Temperature." "In year 21, a reach across the chasm."

U.S. News and World Report: "Who gains in a test-ban treaty: . . . Testing huge bombs, where Russia leads, is out. Testing tactical weapons, where Russia lags, can go on."

There's a test of nuclear hardness planned. Silos will be built in Alaska. Nuclear bombs will be exploded in their vicinity. Instruments will record damage. At last one will know exactly how strong the silos are. The test is called "Arctic Night." The phrase is ominous. It sounds like the end of the world, the final darkness, creeping down from the north. Albert Latter thinks Arctic Night should occur, to prevent that eventuality. If you know how strong your missile silos are, you know how safe you are from attack. You know how accurate the enemy must be to knock you out; you know how accurate you must be to knock him out. With knowledge, you can walk the fine wire of deterrence over the pit of war with more precision. The Kennedy–Dean Rusk–McNamara limited test ban will eliminate Arctic Night. McNamara says the missiles are already safe enough, even if you don't quite know how safe. The Russians only have big missiles, he says; they can't hit thousands of little targets.

Robert S. McNamara, in that dramatic speech, tells the Senate Foreign Relations Committee: "Our missile force is deployed so as to assure that, under any conceivable Soviet first strike, a substantial portion of it would remain in firing condition. Most of the land-based portion of the force has been hardened, as well as dispersed."

Latter: "A few days after the speech Harold Brown went to Congress as the Pentagon's technical guy, to defend McNamara. A few days after that I was in Harold Brown's office."

Halls of the Pentagon, 1963: Photographs of aircraft: B-52s, F-4 fighters. Photographs of soldiers using M-4 rifles. Photographs of Green Beret advisers in the Republic of Vietnam. Little yellow electric carts carrying cardboard boxes, whipping around corners. Woe to the pedestrian. Albert Latter on the third floor, striding through the corridors. His hair is darker, slightly less receding. A wry fury is in his eyes. There are people in the halls here who recognize him, and do not entirely welcome his presence. Schaffer: "Al Latter was a force." Even in 1963 Latter already has a reputation as a man who interferes; a man who shows up, has quiet, knowledgeable words with the mighty, then leaves, and behind him there is suddenly a chaos of new ideas.

Harold Brown, that taciturn, shy, brilliant man, who had earned his doctorate in physics at Columbia at the age of twenty-two, and had become the Pentagon Director of Defense Research and Engineering at thirty-three, has an office on the third floor. A small sign, letters pressed in cardboard: "D.D.R.&E." Dr. Latter to see **Dr. Brown**. Is he expecting you? Yes. A photograph on the wall of a Minuteman I missile, tiered stages, leaving Vandenberg Air Force Base on an orange-white flame. Brown: A young man with a square face and tufted dark eyebrows. A man who likes documents. Documents are piled on a table. Brown will read them tonight. Hello, Albert.

Latter paces the floor. He looks at the curtains on Brown's windows. He stares out into Brown's hall. Arctic Night. Silo hardness. Hardened and dispersed! McNamara thinks our missiles are safe from Soviet attack. They're not.

"I told Harold I thought the test ban treaty would be a disservice to the country."

Harold, McNamara says the Soviets can level New York, but can't pick off all those dispersed missiles. If the Soviets tried it, he thinks, they would be hunting ducks with cannon — lots of noise in the woods, but few birds in the bag. A foolish rationale. All the Soviets have to do is chop those big warheads up into little warheads and send each on its separate way.

Won't work, Albert. Just finished testifying to that effect in Congress. How are you going to get the things off the bus? In years to come Harold Brown will become known among weapons procurement enthusiasts in the Air Force as Dr. No. Won't work, Albert.

I think you're wrong. Yes, you can make little warheads out of big ones. Yes, you can load a handful into your missile. Yes, you can send them each upon their separate ways. We can do it. The Soviets can do it if they choose. It's utterly new. It's never been done. We can do it.

Latter looked hard at Michael. "I went to Rockwell

Autonetics. I knew people in guidance and control there. I asked them, could you do this? They said it was technically feasible."

Latter paced across the room and stared out into the hall. He had been drawing lines and circles on the blackboard, dividing up warheads. Michael copied some into his notebook. Latter seemed about to speak to an invisible person outside, but then he turned and looked towards the city.

"Brown said nothing was going to stop the test ban treaty," Latter said. "But he said: 'If you really think this idea makes sense, I'll put some money into it.' "

Outside, the sunlight glittered on the huge, clear city, and on the Pacific. The wind clattered lines against masts in the blue and white forest of the marina. Latter's eyes were narrowed.

"So he got money into the D.D.R. and E. budget," he said, "and we influenced the Air Force in San Bernardino through various committees... So we introduced the notion of separately aimed warheads. By nineteen sixty-five or sixty-six it was a recognized technology, so you could no longer ignore the threat. The Soviets probably had never even heard of it, or probably didn't care. But it was possible. It was no longer violating a law of nature to say there was a threat."

Everything in the RDA offices on the tenth floor was classified. The filing cabinets, hidden behind sliding wooden doors, were all mounted with combination locks. Michael's visitor's badge read "ESCORT REQUIRED." Latter escorted him back to the elevator. Michael looked around on the way. Secretaries, laughter, typewriter clatter. As they waited by the door Latter murmured: "Lloyd Wilson was very active in the early days of the MX. He's a guy it would have been helpful for you to talk to."

Michael lifted his notebook. "Oh? Where is he these days?"

Latter shook his head slightly as if he had just remembered.

"Unfortunately," he said, "he's dead."

⑥

I

MICHAEL called Neil Buttimer again.

"Any hope for the interview with Bill Crabtree?" he said.

"Colonel Crabtree's pretty busy these days," Buttimer said. "I'll work in it. I'll give you a call."

II

MICHAEL was filling his files with newspaper articles. The New York *Times* reported that the new secretary of Defense, Caspar Weinberger, was on a collision course with the Air Force over deployment of the MX. The Washington *Post* reported that the Office of Technology Assessment had unenthusiastically recommended the deployment of the MX missile in small submarines at sea. The Chicago *Tribune* published a story in which several experts cast doubt on missile accuracy and consequently the need for the MX basing system. Stansfield Turner, former director of the Central Intelligence Agency, wrote an article for the New York *Times* magazine arguing that the shell game MX "would move us in the direction of greater instability." Columnist Russell Baker suggested an alternative to that basing — shuffling the entire Pentagon and 250 phony Pentagons around on the Interstate highway system. The real Pentagon released new reports on Soviet activities all over the globe: Gentlemen, the Soviet Union's arms buildup is apparently continuing. Arms are flowing into Central America. We have reliable information . . .

It was late in the spring. Michael drove south along the coast of California. The enormous freeways flowed like rivers. Elegant housing tracts, as large as cities, blossomed beside lakes of imported water where children laughed in dinghies under red and yellow sails. The older towns — San Clemente, San Juan Capistrano — nestled under their grown-up trees while an ocean breeze carried shreds of featherweight white cloud inland off the water.

Major General John Toomay's house, an unpretentious bungalow, was on the west side of a new road just behind a hill from the sea on the edge of the town of Carlsbad.

Inside, Michael sat in a deep white chair and listened as wind chimes in the window rang in the breeze. Virginia Toomay was in the kitchen making tacos. A bowl of tortilla chips rested on the round, wooden coffee table. John Toomay sat on the couch, his long legs stretched out across the carpet. His hands were large. His white hair was cut to a half an inch from his scalp. His face was big. His speech was almost languorous, a steady rumble that had, instead of just power, a sense of direction.

"I want you to be an expert at this," Toomay said. "It's not difficult."

Michael sat across the table from Toomay. There were the sounds of the chimes and sounds of cooking from the kitchen. It was a comfortable place: a soft carpet, pale furniture, and outside the breeze.

John Toomay was six feet, seven inches tall. Michael watched his face as he talked. It might have been molded in coarse clay by a philosophically casual artist who had heroic inclinations. It was a bold, frank face. Michael smiled. There were no electronic sentries here. He put a little tape recorder on the table. A tiny red light showed that it was working.

"I want you to be an expert," Toomay said. "There are really only a few things. The fundamental view that I have is that the Soviets haven't attacked, and we don't want them to attack, and the way to do that is to be absoluely sure that you have a set of systems in which there's not a single niche where they can jump through and see themselves as being better off at some stage of a nuclear war, you name the stage."

Toomay was slumped far down in the couch, his legs stretched out for a meter or so in front of him. For four years just after World War II he had played professional basketball: the Chicago Stags; the Washington Capitals; the Baltimore Bullets; the Denver Nuggets. That had been his main civilian occupation. Korea had dragged him back into the Air Force and there he had stayed. Now he talked like a coach explaining the philosophical foundation of the fast break or of the stall on the ledge of the two-point lead.

"In your mind you visualize this whole hierarchy of things that can happen, from one nuclear burst somewhere all the way up to total holocaust, and in each one of those situations the Soviets should look at that and say, if I ever did that I'd really be stupid because the United States can do this.

"I think I can explain to you what the threat is, because there is only one threat, and I think that James Schlesinger and Harold Brown and others keep expressing it over and over again, and I don't know if anybody listens. But the threat is for a disarming first strike by the Soviet Union. The argument is that it isn't enough just to have one system which you think is going to survive. You really need three. Three came about as a happenstance for other reasons, but once three got there people started thinking up reasons for having three, and now it appears that three is a very good number to have. The argument is that one of them can fail because of some unforeseen technical glitch because none of these weapons has ever been fired in the way it would be fired if anything happened. And one could be subject to unforeseen technological surprise — in other words, to unforeseen defeat by the Soviets. The philosophy has always been to have separate modes of survivability of our nuclear weapons; independent modes of survivability so the Soviets can't develop one piece of technology which will bridge over from one leg of what is called the triad to the other. That's why the sea, the land, and airborne all represent separate survivability modes.

"What it really is, as I see it, is just making damn sure that as the Soviets do these calculations, if they do them, they look at the problem and they say, 'Hell, there's no chance, it's not going to work, forget it!' "

In the bright California home, the wind chimes rang like the musical interpretation of a brook. Michael leaned back in the chair. He glanced at his tape recorder. He could see the tiny reels turning. Toomay rolled on.

"People argue. The unilateral disarmament guy would say it's all intolerable because civilization will be obliterated by a massive nuclear exchange: it's better to have a civilization even if the Soviets dominate it. The mutual assured destruction guys simply say that compared to the dangers of a nuclear exchange, the dangers inherent in Soviet adventurism and other things are really not that great, so what we should do is not worry if we've got a couple of submarines that are working because we know we can get their cities. The guys who believe that we need to show our strength and our will to use our forces believe that the Soviets can stand an awful lot in the way of losses in order to reach their objectives, and they point at ten million starved Kulaks to collectivize farms, purges that result in millions of dead, twenty million losses in World War II and numbers like that, which we could not tolerate in this country but which they are ready to take in stride as proven by their past acts."

At last Michael found a pause in this steady monologue in which to plunge a question: "The question of how the MX would be used — I was reading a strategic planner who said it could be used in specific limited-war ways — that you could think of a small nuclear war because of MX — doesn't that make the war more likely?"

"I," said Toomay, and shifted in the couch. "— God damn it, I — it's hard for me to accept the reality of that except in light of potential Soviet calculations. This idea of carrying on a nuclear war after a massive nuclear exchange — bah, crap, baloney. The people are going to be hunting for their own kin, and they're going to be looking out after themselves, and it's going to be hard to generate much interest at all in what goes on on a national basis. Those guys! It's one thing to talk about the perception of our capability on the part of the Soviets, and I'm all for that. But this idea that you're really getting down in there and thinking about fighting that way is just bullshit. It's just utter bullshit. It's incredible. When the people jump up in these audiences and say you guys are crazy as hell, that's what they're complaining about.

"I can't handle the idea that we're building MX so that after fifty million people are killed we can still do something meaningful to the Soviets, say, 'Ah, you bastards, we've showed you!' That doesn't do a thing for me; that just depresses me.

"Because of course the objective is to stop that from happening, not to win it after it happens. But the argument is in order to stop it from happening you have to worry about how you would win it if it happens, because the Soviets have to see that you would win it if it happened."

That was as complicated a train of reason as Toomay would allow himself. He stopped. Virginia Toomay brought each of them a glass of lemonade. Then the steady, unemotional voice went on.

"I've been thinking about this damn subject for fifteen years, and I finally settled on a way of looking at it that I can handle. I consider myself a practical guy who thinks freedom is extremely important, perhaps more important than any individual's life, but who doesn't think that making wars is a very useful thing to do."

Michael searched the general's face. It was momentarily at rest, utterly calm. The tape recorder ran on, scrupulously taking down silence. It's light looked like a little red eye. Toomay gazed thoughtfully at Michael. Michael glanced down at the pages of his notebook.

"Why don't you stay for dinner?" Toomay asked. Michael looked up, startled, embarrassed. "Just tacos," Toomay said. "But they'll be good."

48

By the time Michael left the Toomays it was night on the freeway, a wilderness of white and red lights and noise. Blazing gas station and motel signs; shopping malls, subdivisions, lighted crosses, white lines on tarmac, the river of cars, lights in windows, the deep, safe sky.

7

SUBMARINE LAUNCHED BALLISTIC MISSILES
United States: 656
Soviet Union: 989

I

THERE were two filing cabinets in Dr. Herbert York's office in San Diego. They were labled "ABM to NPT," and "Nuclear Accidents to Vladivostok." Michael sat quietly, his tape recorder running. York, a physicist who had been one of the early developers of the hydrogen bomb, had later written a book called *Race to Oblivion* about the arms race. He had been the first in the powerful position later occupied by both Harold Brown and Bill Perry: D.D.R.&E. He was a piece of the military establishment that had spun off to become a spark of dissent.

In a photograph on the back of his book, York looked like an enormous frog with tiny eyes. In person he was indeed large, but his eyes were gently, kindly trusting.

Michael asked:

"Is there any type of personality that tends to go from

physical science into politics, like weapons planning?"

York smiled.

"There are broad generalizations, but I don't want to make them."

Michael waited. York went on reluctantly:

"A lot of scientists and engineers do quite badly when they get into the realm of trying to make policy decisions," he said. "They tend to look for single causes for events, and when they find one they stop looking for more. In the political world the norm is that any particular event usually has lots of causes, and not just a single one. These causes are in quite a few different dimensions, so that the whole cause-and-effect chain is different in political life than it is in scientific life, and many scientists have no appreciation of that difference at all."

Michael said:

"Is there something that happens to a person, in their own — self-image when they suddenly become involved in this issue?"

"They very seldom suddenly become involved," York said with a smile. "If you're studying quasars in graduate school, say, you get hired at Los Alamos to do calculations that have something to do with building hydrogen bombs, and in the course of that you start hearing briefings and discussions about the necessity of having hydrogen bombs. You may even come in and have some questions and some doubts: Why are we building these? So it may just be your group leader who explains to you the facts about the Russians and what we have to have. I think characteristically it happens very slowly.

"Briefings about the threat are among the principal elements of life in any military industry. You're just always being briefed on the threat. Then in the world at large you're always hearing accusations that the threat is exaggerated. So you both learn that the situation is desperate and needs fixing, and you also learn that there are a lot of people out there who just don't understand. So you're subject to both of those ideas and they're constantly being reinforced."

On one of the steel shelves in York's office was the small set of models of United States and Soviet missiles, that Michael saw everywhere: a comparison of cigars. There were books: *Clausewitz on War, Race to Oblivion, The Aerospace Corporation 1960 to 1980.* On the wall were photographs of York with famous figures, including President John F. Kennedy.

"The basic problem that we're all faced with," York said, "including you, is that there is a real threat, but there also is a strong pressure to greatly exaggerate it on the part of insiders. The whole process selects people who believe that technology is the answer. Because almost the only way you get to be an insider is by believing that there is a serious military threat which can best be handled by pushing ahead with your own military technology."

York smiled. An expression of faint sadness dominated his huge figure. "And a few of them, or a few of us — I'm not sure which to say — have come to realize that it's not as good an answer as we once thought it was."

II

Michael stood at a bus stop outside Washington, D.C., watching snow accumulate on his shoes. In all the little fenced yards of the homes around him the snow lay three or four inches deep. The yards were unsullied by footprints, as if the residents who had already left for work had cherished the morning's freshness and had tiptoed out, leaving the purity for their children. The bare trees were frosted, the evergreens had been charmed. The falling snow muffled even the roar of the traffic that tried to defy it.

The bus came. Michael took it to a subway stop. He took the subway to the Pentagon. Up escalators, up ramps, and — he got lost. He was looking for 3E130: third floor, E ring, room 130. But the Pentagon engulfed him and spun him around. He wandered down hall after hall. He passed battle scenes hung on yellow walls: burning ships, mortar fire, blackened faces shouting. With burning feet, he found himself picking his way through rubble. Smoke and dust was in the air. He asked a man on one of those confounded yellow carts where the E ring was. The man pointed back the way Michael had come. He seemed to be looking up at the sun. It was soft and haloed, the way it gets before a storm. The E ring was the outer ring. Third floor. He was looking for a road; he could see only pathways through shafts of broken concrete. Everything tilted and seemed about to fall and crush him; but he was accustomed to fear. He went up a ramp and a stairway: Room 3E1013; 3E130 must be on the other side of the building. He walked and walked. From a leaning windowless structure a hunk of concrete hung by a single strand of a reinforcing bar; it swung very slowly, like a pendulum on a run down clock. 3E655. He walked and walked. There were permanent shadow silhouettes of people on the concrete walls. The shadows were the places where the people had been. They were negatives, where the bodies of the people had protected the wall from the heat. The people were gone. 3E510. A little yellow cart carrying cardboard boxes beeped at him and turned a corner. The debris was thinning out now; there was more dust and more fire; more charcoal where fire had been. More silence. 3E410. Now the land was almost flat; windblown; humps in a field. 3E229. Michael climbed a little hill of pulverized earth. In the smoke there was no horizon. He looked down on a shallow circular pit with a faintly green rim. Within it there was no steam, no stirred dust, no smoke. It was wholly without life. He shouted. There was no echo. He turned and ran.

3E130. A small cardboard sign. Michael had almost circumnavigated the entire Pentagon. It appeared that the rooms numbered in the thousands were separated by one hallway from those in the hundreds. He was just yards from where he started. His feet were hot. He could smell his sweat.

He left the bland yellow hallways and entered the raucous office occupied by Seymour Zeiberg. An enormous mobile hung above the vast desk. The mobile was of the cartoon beagle, Snoopy, riding a red doghouse in pursuit of a model biplane. On a nearby table stood a spray can, labeled in large letters: "BULLSHIT RE-PELLENT."

Michael carried the same notebook he had used in his brief interview with Gary Aubert, the man with the map and the pins, on the opposite coast. In it were Aubert's remarks on Zeiberg. There were few men more influential in the Pentagon, Aubert had said:

"You have to enjoy a position of power to do that job," he had said. "That's hardball. Here's a chance to do something to go down in history, hopefully positively. Weapons development policy: that's the source of the Ganges."

Michael sat down at a table at the source of the Ganges where both he and Zeiberg would be out of reach of the can of repellent.

Zeiberg was a stocky man with a crewcut. He looked comfortable, powerful, decidedly not hungry, and, although he sat still for the interview, he seemed to be in deliberate motion. He seemed to have momentum. Michael didn't see him breaking through barriers to win his battles. He saw him crashing through floors, like a dropped anvil, getting where he wanted to go.

Michael asked his questions, watching the man. Zeiberg answered with a patient amusement in his eyes, backed by condescension. Michael smiled and nodded as he spoke. Michael's eyes were cold. Zeiberg's were cautious. Right now, Zeiberg said patiently, he was caught in the gulf between administrations: the infamous MX basing mode was up for review by the new Reagan administration, and he was supplying information to one of an apparently endless succession of committees that were studying the options for a new decision. He discussed it as if he was personally remote from the outcome.

"I have to put the material in a form that ties it together for presentation," he said. "Organizing the stories on the various alternatives that have been seriously considered in some standard format so it's easy for the new players to assimilate it."

Zeiberg mentioned a journalist who had recently attacked the MX missile in *Rolling Stone* magazine. "Some new face," he said. Michael smiled. He asked:

"In the past decade or so from what you have seen of the years in the interim — these people coming in to make a decision — there really isn't any new material for them, is there?"

Zeiberg said abruptly:

"There hasn't been anything fundamentally new for years."

"Do you think that — will they have to decide to continue what you have currently planned to — something similar to the current design?"

"It will have to be. I see a very digital decision, very binary. Either something like what we have, with some minor changes, or dropping the idea of a new, survivable land based missile."

"Are they strictly looking at the basing mode — is that correct? Not at other conceptual —"

Zeiberg interrupted with a little twitch of impatience.

"They were told to accept the need for a missile of the military capability that the MX has."

Michael began asking questions about history, and a little of Zeiberg's past emerged.

Back in the second half of the 1960s, he said, after Latter had made his brief remarks on MIRVs to Harold Brown, Zeiberg had worked with John Toomay, among others, out in San Bernardino, near Norton Air Force Base. Toomay was a colonel. They worked on missiles. What they were doing had nothing to do with the threat as invented by Albert Latter; they were working to nullify the threat of Soviet defense.

The concern at the time was that the Soviet Union was building a huge new anti-ballistic missile system that could knock off U.S. missiles as they arrived. This defense was an offensive threat because strategists figured that if the Soviet Union could stop incoming missiles then they could attack the United States without worrying about retaliation. So one of the hot new fields of research was in the development of devices to get warheads through this flak. Zeiberg was employed by the Aerospace Corporation in a program called Advanced Ballistic Re-Entry Systems. It was known as ABRES. In his briefcase Michael had a list of other projects that shared the same acronymous penetration aids zoo with ABRES: ABALL, ECM, Active ECM, OPADEC, Drads, LORV, MBRV, REX, REST, WAC and Sleigh Ride, which had been previously called Reindeer. The devices Zeiberg and Toomay worked on were called Penetration Aids.

"I'm a New York boy," Zeiberg said. "I started out as a mechanical engineer. I went to a technical high school to train myself to go into the field. I went to City College of New York for my undergraduate work. And New York University for graduate work." He developed a more casual look of amusement, reminiscing.

"As I started taking fluid mechanics in college I developed an interest there and then generalized as I went deeper into it. I decided to be an academic. I stayed on at CCNY teaching while I was doing graduate work at NYU. I stuck with it a year after I finished my doctorate work, when I lost motivation and also acquired some greater expenses; my first child happened to have been born. So I went to work for General Applied Science Labs, in fluid mechanics and aerodynamics. That was in 'sixty-two."

Zeiberg leaned back and folded his hands over his stomach. He went on: "By then the business of observing reentry vehicles with ground based radar and optical systems and airbone instrumentation, all aimed at supporting Anti-Ballistic Missile programs, was getting very very intense. And I got into the business of modeling the aerodynamics process around the reentry objects. And that directed me into the question of how do you design these things, and then how do you use them, so I started interacting more with the design community and the systems community and developed an interest in that aspect, and was recruited away by the Aerospace Corporation in nineteen sixty-five. Then, having gotten into that, I directed my activity to the broader aspects of the design of weapons systems."

Zeiberg's office was warm. His voice thumped along. Michael was going back.

Norton Air Force Base, 1967:
The air over San Bernardino is still clear most days. Zeiberg has a house on the outskirts of town. He has a comfortable home with water, sewer, power, and schools, but at work he is out there with the Indian scouts. If you have always wanted to know the way air flows around something the shape of a football as opposed to something the shape of a pencil, when both objects are going 15,000 feet per second, and you can talk up a need for that knowledge, the money is there. If you feel like finding out how heat resistant a mixture of aluminum, titanium, steel, and, what the hell, mustard, is, and you think maybe it will help a warhead survive reentry, the money is there. More to the point, if you need to know exactly how hot a Teflon sheath on a conical warhead can get in atmospheric reentry before it ablates away and contributes to the reentry vehicle's fiery wake, you don't just sketch it all out on paper and ship it up the chain of command, you get on a little plane in San Bernardino, nicknamed Pinky's Airline, you catch the redeye to Washington, you argue your case at the Institute for Defense Analysis and in the Pentagon, you get the funds, you go down to Green River, Wyoming, and you shoot an Athena missile down the White Sands Range and find out. It is applied science in an unlimited laboratory. It is almost incidental that these dramatic and fascinating issues are being examined in the context of the waging of atomic war. Except that if you ever get bored with this fabulous undertaking the threat assess-

ment people are always around to remind you of its meaning.

In Zeiberg's office Michael flipped backward through his notebook. Another of Zeiberg's colleagues had said: "We used to get weekly briefings by Aerospace people on developments. At Aerospace, we already felt motivated. Our insights into Soviet developments urged us on even more." Zeiberg talked on. Michael imagined the endless briefing.

A slender, polished young man in an immaculate suit. Viewgraphs. A slide projector showing fuzzy photographs. A large room. Civilians. Blue-suited soldiers. Seymour Zeiberg is here. His counterpart from the Air Force, Colonel John Toomay, is here.

Gentlemen, the Soviets have been detected installing the final panels in their new radar systems outside Leningrad. We have detected a new array added to the Soviet installation near Moscow. Gentlemen, it has come to our attention that the new Soviet missile, the ABM-1 Galosh, can carry one nuclear payload with a slant range of two hundred miles. Gentlemen, we suspect that the new radar system may be related to the development of the Soviet ABM. It has been estimated that if the Soviet radar is of the appropriate type, it will be able to discriminate small differences in reentry vehicles at altitudes of greater than one hundred miles. Gentlemen . . .

Zeiberg: "The perspective we were working on was catch up perspective. They were off building these immense radars. We were really worried they would deploy something . . ."

Good lord — they'll be able to pick out differences in the RVs at a hundred miles up? Back to the boards,

John, Sy, Dick, Gary. That damn decoy has to look more real. If the ABM system that the Soviets seem to be building can spot incoming RVs and knock them off before they came close enough to spread their venom over the towns and villages of the U.S.S.R., then we have to confuse the ABM system. We have to send those interceptor rockets dashing off to blast shreds of tin out of the sky, while the warheads escape. We have to fill that searching radar eye with hundreds of identical blips, only one of which is deadly. We have to tone down the radar image of the warhead while we tone up the reflection of the little tin pyramids or tumbling metal rods, or self-inflating mylar balloons that pop out of the nosecone with it.

Sy, what about this: let's call it POOFF. Think it'll work? Let's go down to Green River and put it on Athena and fire it downrange. Well, hell, let's just forget about all these decoys and just put a bunch of warheads in there and saturate the Soviets. Everything we throw at them will look like a warhead because it will BE a warhead. MIRV!

The pressure: If we don't come up with solutions, if suddenly the ABM system becomes a true shield, my God! The consequences! Gary Aubert: "You can't be frivolous about it. It's the most important thing going."

Zeiberg has the technical knowledge and the broad view. He has quickly become a technical manager. He runs people and programs. Remarks familiar to those who know Seymour Zeiberg: "Optimum, schmoptimum. All I care about it is that it works." "He wouldn't know a decoy if it hit him in the ass."

Zeiberg has a small grenade he keeps on his desk. No one knows for sure whether or not it is loaded. During briefings from subordinates Zeiberg may start playing with the pin. Briefers get the message. They cut it short.

Zeiberg also has a small replica of the Rodin statue *The Thinker*. He tends to place this in the middle of his desk or conference table, and during meetings refers to it rhetorically for advice. One day Zeiberg has been listening to a discussion about penetration by men who work for him. Perhaps the men are new players. Zeiberg watches them with a kind condescension. Near the end of the briefing, Zeiberg interrupts the men, and says, "Let's see what the thinker has to say about that."

He lifts the little statue. Underneath is a pile of fake dog manure.

The marvelous work continues: thirty-degree angle of impact with the atmosphere. Mach 6 speed at 400,000 feet. If your decoy's too light the atmosphere will sort the real thing from all the dummies as they burn up. With a quick reaction intercepting missile he can wait and then slap it. Get the decoy down to 50,000 feet, and then it'll be too late. What will diminish the beacon of the plasma sheath? Chrysler's been working on bleeding ionized sodium through the pores of an RV. Let's get the data! What if you make up for the lack of mass of the decoy with a little motor: FIRE it down! Put a lot of corners on the decoy and none on the RV; they'll look alike in the scope. Spoof the radar with a maze of decoys. Company in Ann Arbor makes a coating material that dampens reflection in one frequency or many. Mix it with plastic and you have an ablating, absorbing cover. OK, put it on an Atlas and shoot it down to Kwajalein. Stick it on an Athena and shoot it down to White Sands!

Gentlemen, we have reliable information that the Soviet Union is developing ultraviolet sensors that may be able to accurately measure the ion sheath of reentry vehicles at greater range. Gentlemen, it is possible that

Soviet radar systems will be able to determine mass differences between decoys and RVs at a less than twenty-to-one ratio. Gentlemen . . .

With Snoopy dogfighting over his left shoulder, Zeiberg sat back in his chair, comfortable in his memories. Michael's eyes were bright. He watched Zeiberg with an intensity that was almost like affection. The excitement of the work! The camaraderie! The challenge! Thinking up what had never before been thought to solve the ultimate problem that had never before been encountered. The magic of invention!

"We were at a stage of the fast part of the learning curve," Zeiberg said. "We had responsibility for formulation and management of a lot of programs, and then acting as the independent technical auditor for the Air Force on a lot of this." Then he grinned, a real smile, without condescension: "Every day was a new and exciting venture."

"Yeah," Michael said, with zest that seemed to surprise him. "I can see how it must have been."

Zeiberg had another appointment. He shook hands pleasantly. His hands were large. Michael walked quickly down the stairs. He passed through the halls of the Pentagon without trouble. He walked rapidly down a ramp and out among the stores of the concourse entrance. Fannie May Candies. Walgreen's Drugs. Near the escalator to the subway was a display set up by the Army. In his headlong rush, still flinging Teflon-coated reentry vehicles through space, Michael almost passed it blindly.

But it stopped him. "Have you heard about . . .?" the display said incoherently. "Have you heard? Have you?" It showed pictures of green helicopters and men in camouflaged helmets. On top of the display was one word: "Threat." On a companion display was the same word. "Threat." Under it was an outline map of the world. In what looked like a half-dozen places or more — Central America, South America, Western Europe, Eastern Europe, the Middle East, Southeast Asia, India — large round red lights glowed in the map. Michael stood and looked at them. They reminded him of something. A montage of photos next to them showed tanks, aircraft, soldiers, bridges, faces. He examined it. Vehicles in the Army used to be uniformly green. Most of the vehicles shown here were painted with brown and gray desert camouflage. Michael looked at the red circles on the world again. They looked like the blots on an old civil defense map of the United States he had seen recently: they looked like the red zones that symbolized the areas likely to be obliterated in an all-out nuclear attack.

The threat. The threat. Gentlemen, gentlemen. Your attention please. The Soviets are learning. The Soviets are catching up. The Soviets are pulling ahead. Michael looked at his watch. The subway was late. He waited for his ride in the vast concrete cavern of the subway station. The train came whistling down the tunnel, blowing air ahead of it like a storm.

$$\bigotimes$$

SUBMARINE LAUNCHED BALLISTIC MISSILE WARHEADS

| *United States:* | *5120* |
| *Soviet Union:* | *1309* |

I

MICHAEL stayed in the home of relatives just outside of Washington. The home was near the Bethesda Rescue Station. Michael had a pleasant room at the top of a three-story building. The room looked out into the tops of bare trees. The rescue station was always busy. Regularly, day or night, sirens would break out just down the street, howling of catastrophe. Each time it happened, Michael, whose home was in a farm valley in Montana where sirens rang once or twice a month, would look up and stare out through the branches at red lights flashing against buildings, dread in his eyes.

He had several tapes with him of previous interviews. At night he would listen to them and transcribe or make notes. On the night of that snowy day he listened to recordings of John Toomay. The sound of the retired general's voice came strangely into the cold room. Disembodied, it seemed warm. In the background of the tape Michael could hear the sound of the chimes singing of the California breeze.

"Zeiberg's a younger guy than I am by a considerable amount," Toomay's voice said. "He'd just gotten his degree, worked for a short time for a rather narrow technical group, and then he came out to San Bernardino and joined Aerospace Corporation. At first he was just part of the guys who were examining the flow over the reentry body, you know, and whether it's turbulent, and all this stuff, and then he rapidly rose to a point where he was worrying about what kind of penetration systems we should have and what kind of reentry vehicles. Well, suddenly, you see, his technical work brought him into an arena. I don't know how much he thought about the megadeath type of thing, till he got into the arena where he was actually thinking about and planning penetration systems."

On the tape, Toomay paused. The room was silent. Then the tape went on. Disembodied, Toomay's voice was just like him: large, steady, relaxed.

"You see, that's the same thing happened to me. I went out and got my engineering degree and I came back and somebody said, 'YOU, go be in ballistic missile defense, radars.' And I did, and for a while I just worked on radars, and then I began to see what these radars can do, and then I began to look at the global ramifications of what I was working on and then you have to make peace with yourself, and say well, this ballistic missile defense is rational; we're going to save millions. And so away you go.

"Well, Zeiberg must have thought the same thing. That's the moment when he had to come face to face with the idea that his science and engineering was at least involved with his global view of the world. Yes, that would be a time in his life when that happened."

Michael's voice appeared on the tape. In the cold room in Bethesda Michael grimaced. Outside cars were hissing past on the road. Michael's voice, sounding unnaturally cheerful, said:

"Is there always — Is there sort of a bunch of young guys going through this in various — in the military and engineering fields?"

"Exactly," Toomay said. "All the time."

"How do they communicate the issues that they come up against? Do they — is there a dialogue going on?" Listening, Michael drummed his fingers on the desk.

"I think there is," Toomay replied. "I think it is not formal. Now, if you're a political scientist or something, you start out in school thinking about these things at the outset, whereas most engineers read *Time* magazine or something and that's sort of an aside, and then all of a sudden as a result of moving up in their discipline they come face to face with that. It's a little different if you came up as a nuclear weapon designer — those guys had to think that out ahead of time, to my way of thinking, because they're into megadeaths as soon as they get to graduate work. But a lot of these other guys, you know, they're just following their discipline and maybe they came into the military for many other reasons."

There was a little pause. The sound-seeking microphone in the recorder found the chimes. It drew them up out of the background. When Toomay's voice returned it blasted them back into insignificance.

"A lot of people join the military because they like to know where they stand. Socially and psychologically.

There's never any doubt where you stand. My brothers have accused me, you know. You! I keep telling them I'm not a militarist, I just showed up for a war, and showed up for another war, and then inertia carried me on. They said, 'That's crap. You were born for the military. You always want to know; everything is mechanistic to you. You always want to know where you stand in relation to everybody else, and what's going to happen next, and you need to be in this very well-defined, graduated, structured environment.'

"When I thought about that, and this is just in the last few months, maybe they're right.

"So, anyway, these guys gravitate to the military, some of them, for entirely different reasons, then all of a sudden they find themselves having to make these various decisions. Now, I view these pilots as having to do that same thing. When they first take up flying all they want to do is fly, and it's adventurous to be a fighter pilot, they don't think about actually being in a war, I don't think. Then when you become a bomber pilot and get sent to the Strategic Air Command, and right away you've got to face up to the fact of what you're going to be doing. I think most of them, you know, peace is our profession. Those slogans — they sort of go for those. They never question those during their lives unless some event occurs which causes them a lot of trauma. Vietnam was an event like that. It started a lot of people asking questions. But if you notice, all the Air Force and Navy pilots — old guys, young guys, middle-aged guys — they all went ahead and fought. So you can see that even those are acid decisions, most people don't have that much trouble making them. Figuring out a reason why they're killing people.

"I think that a lot of these people in these positions where the megadeaths are there really do their rationalizing on a much lower level. It's hard to describe, but — But when I was a lieutenant colonel in ABRES I was working opposite Zeiberg, you know, we were worrying about these penetration systems, and we had about a hundred and fifty million a year we were spending. To say that is really an exaggeration; you know many echelons decide that it's going to be that amount of money.

"The thing was, I had a young officer come in. He was a real dissident. He was just an impossible guy, but — the thing I remember was, I used to counsel him more than I counseled my children, which really was outrageous to me, because the guy didn't deserve counseling. Anyway, he came in there one day, and I was a lieutenant colonel and we were working on these technology programs; he was part of it, and he said to me, 'You know, being in the Air Force is really different, when you have POWER, like YOU have.'

"And I couldn't understand what the hell he was talking about. I thought of myself as a member of a team who was doing his best to organize work. I had never even thought in terms of my having any power."

II

As Michael worked into the night he scattered papers around on desk tops and the floor. The Report of the General Accounting Office on Basing the MX Missile. Transcript of Dr. Seymour Zeiberg. MX notebook II. MX notebook III. MX notebook IV. MX notebook VII. Telephone conversation, General John Hepfer. The Aerospace Company, 1960–1980. A newspaper clipping of

Alexander Haig's views: "I have described the Soviet threat as relentless. That is a very considered term. . . . In a few years we Americans must, indeed, we will, one way or another — make a decision as to whether or not we will continue to seek a world order hospitable to the Christian-Judeo values and interests of today or to abrogate that order to values and interests distinctly different from our own."

Michael listened to tapes. He flipped through the papers. He wrote on the clip of Haig: "Inevitability of war." He threw Haig off into one corner. He looked up notes on telephone conversations. He got up and left the debris and stared out the window.

Norton Air Force Base. The late 1960s. Several of the men on his list had passed through Norton Air Force Base around the end of the 1960s for one reason or another. There were Zeiberg and Toomay, sweating over penetration aids. There was John Hepfer, then a lieutenant colonel, reporting in in June 1967 to work on Minuteman II and III guidance. There was Captain Bill Crabtree, dropping in from 1969 to 1971 to work for an Air Force general people called The Silver Fox on maneuvering reentry vehicles. Crabtree had been working for Jasper Welch in Los Angeles at the Air Force West Coast Study Facility. Marvin Atkins, who had left a twelve-year career with the Air Force in 1964 to go to work for the Avco Corporation in Pennsylvania, showed up a time or two to get briefed, give briefings, and receive instructions on a warhead project he directed.

And there was Albert Latter, drifting down from Los Angeles for Science Advisory Board meetings or consultation, making unexpected suggestions about, say, putting missiles in canals, or running them about between garages in the desert, proposals you had to listen to with care because though you half thought they were crazy you also half thought that Al Latter knew more about this stuff than you did and you KNEW it was entirely likely that Latter had been making the same suggestions to more important ears than yours so someday these ideas would come drifting down from on high.

Norton Air Force Base: desert air, palm trees, an abundance of lawn, surrounded by bleak little windblown huts, homes, bars and pizza parlors, and lots full of sandy bushes. Cars and messages running back and forth between the Ballistic Missile Division offices on the base and the long, low building a few hundred yards to the south, on the corner of Tippecanoe and Waterman: the Aerospace Corporation facility.

Norton: What could be more fun than sharing friendship and vital national secrets, referred to only obscurely, with complete safety, over a desert enchilada at Lupe's in the hot evening sun? Or getting pleasantly elevated among the candles and the elegantly dusty wine bottles at Pitrozello's Del Fino de Oro restaurant after some successful Atlas or Athena shot that has culminated months of twelve-hours-a-day work?

Toomay:

"Nobody told you how to think but they forced you to think logically. And pretty soon you ask yourself this question about deterrence, and you say are these guys right? Could they possibly be wrong or are they right? Then you think through with your own brain the process of going through that and saying 'How does all this work — how do we keep the Soviets at bay?' And you finally fall into the rationale — because it has been thought out by smart people. I felt Zeiberg was the same way. We believed in the doctrine of deterrence by strategic superiority then. It was fading then. Strategic superiority had been demonstrated in sixty-two in the Cuban crisis, so the doctrine was uttered in sort of a logical way by McNamara. We just worried about being able to deliver that capability and being able to know the Russians knew it. We felt comfortable about what we were doing.

"Nobody ever said to themselves: 'Let's go out there and nuke the Russians and turn them into rubble'; and nobody ever said we can't stand this buildup because eventually it's going to cause us both to be obliterated from the earth. We all hoped that neither of those two things would happen. Nobody ever said, 'What the hell are we doing this for?'

"We sat around and talked about deterrence a lot, but never questioned its fundamental precepts, or if we did it was only in passing. We had to talk about it a lot, because we had to put together an algorithm for allocating funds, and the algorithm was if a system is both strategic and tactical then it gets 1.0 for its value to the mission; if it's strategic it gets more than half because strategic is the big umbrella. That kind of thing.

"You get so wrapped up in the development steps; the test flights, and they work, and they don't work and you have to do another one and all that, you really don't notice when the deployment decision is made or something. It's made by some guys in Washington, and you're plowing along working hard to get the thing done.

"What I liked about it, we had a whole lot of very powerful intellectual interactions with a whole lot of intelligent, well-educated people, arguing the issue. It was almost like the good old college bull sessions where you argued about is there a God, and if he takes care of each person or what, and if the universe is finite then what is there beyond it. Those kinds of things, but narrowed down to issues of deterrence and what the technology will allow. Those kinds of interactions. You were always preparing things either for flight test or for budget review; you would go back east to Washington, get on the red-eye, give a briefing up through the channels, and get the money. That was the big objective. We did that a lot.

"Most people who work on those things at the level that we were working on them believe fundamentally in the truth of the approach, and if they don't in this country you're free to make another choice.

"That didn't happen. We didn't have people doing that."

The tape ended. Michael turned it over. The past was brighter than this strange night.

Norton Air Force Base: What could be more bracing than visits from anxious or bombastic men from Washington, from Air Force Systems Command or from D.D.R.&E.: Here comes Lloyd Wilson blowing into town like a thunderstorm, in the position that Seymour Zeiberg himself will have much later. Better grab him while you can, he's the source of the Ganges.

Toomay's memory: Here's this new idea, Lloyd. Incredible! We think it'll work like this, Lloyd. Incredible! We need forty-five million for it, Lloyd. Incredible! And when I say incredible, I mean HORSESHIT!

Toomay on tape: "Oh, yeah. Lloyd Wilson. He was truly a character. He could drink gin from a tumbler. He wound up getting shot."

Norton Air Force Base: Baseball games and blazing

summer days; Good old General Bob Duffy laughing with the boys. Norton Air Force Base: Stick it on an Athena at Green River and shoot it down to White Sands and watch it fly!

Good old Bob Duffy. They call him Duff. The stocky young general with the pipe and all that enthusiasm. Who could have thought what was going to happen to him at White Sands!

Norton Air Force Base. 1967. John Hepfer arrives in June. New Chief of the Guidance and Control Division. Minuteman II not exactly getting where they're pointing it. Maintenance on those arrays of gyroscopes and accelerometers and computers is eating them up. Entire squadrons of Minuteman missiles are going down at once. Here's your budget for the year, colonel: one hundred twenty million dollars. Here's what you're spending per month, colonel: twenty million dollars. Hold it! These are the same contractors who are working on our new systems for Minuteman III? Yup. OK. For Minuteman III we're going to get another bid.

Norton Air Force Base. Hey, that Minuteman II penetration system sure doesn't mix up the warheads and the decoy material. Let's get a team on it. Toomay! You're team leader, Air Force. Zeiberg! You're team leader, Aerospace. Lloyd Wilson doesn't like spending much money on this, so keep it down.

Outside Michael's room there was action at the rescue squad. Doors slid open. Men darted out. A large red and white ambulance slid into the street. Red lights went on. But in the depth of the night, with no traffic, the siren was mute. The whole thing seemed to be done in pantomime, a rehearsal without the music. The lights flashed, and the ambulance raced away in an odd silence. Michael looked at his interview with Zeiberg.

"Lloyd was annoyed that the penetration system was in trouble and didn't work, and he was inclined to think the Air Force should just go spend the money on MIRVing. So Lloyd was very lukewarm to the whole idea. The general figured out a strategy. He just told Lloyd what was needed for the first year's funding, and it was late in that fiscal year. And for some reason Wilson thought we could do this whole thing for a few hundred thousand dollars. It was really a few million dollars. So Lloyd finally gave up. He said, 'Ah, it's not big enough to worry about,' so we went ahead and did it. Then when it worked he was very proud that he did it. That had a particularly important bearing on both my career and Toomay's career."

Norton Air Force Base. 1967. John Toomay. A man not known for reticence. He can hide his point of view about as easily as he can hide his frame. Did you hear what Toomay did Tuesday? Listen, here was Toomay, Lieutenant Colonel Toomay. Big meeting, Three Star up front. Three Star's talking about chaff. Three Star's a little confused, just a little. So Toomay says, courteously: Sir, you're ALL FUCKED UP!

John Toomay. General Duffy is at home with his family. There's a knock at the door. The youngest child runs to open it. Her name is Patsy. She's four years old. She pulls back the door and she looks up. She looks up and up. She looks up still farther. She turns around.

"Daddy," she says, "it's a gi'nt."

Norton Air Force Base. 1967. Colonel John Hepfer. Did you hear what happened after Hepfer told the contractor he was looking for another bid? The company had fourteen thousand people working on Minuteman. Hey, in a year they cut back to seven thousand people working on it, and then they underran the contract by twenty-five million. Three RVs on each Minuteman. Three MIRVed RVs. Autonetics'll do it. How many RVs will they get on that new missile, the WS120 they're talking about. A hundred-inch diameter monster! Fifteen?

Hey, did you hear about what Hepfer was working on before this operation? Electronic Fence, Vietnam. Drop sensors all over the jungle along trails. Hang them up in the trees. Supposed to pick up the sounds of passing Cong. Find out where the trucks park at night. Hit 'em with F-4s. Well, hell, most of the people coming down the trail walked or rode bicycles. We wanted to count them. So we wanted to listen to 'em. They don't sing cadence. So this project Hepfer was on decided to make tiny bombs that'd pop underfoot. Some army guy figured out the specs. Couldn't hurt or you wouldn't walk on them. Just go pop, so the sensors could hear. "Button bomblets." Made them, put them on the trail. So the Cong got out the brooms and swept the trail. Listen, old Hepfer's no dummy. It was kind of a ragged project. He didn't know what they wanted him to do. Well, he'd just graduated from a systems analysis course, so he analyzed something. He analyzed the cost of killing a truck with all this stuff and an F-4. Came out to about eighteen million dollars per truck. So they sent him here. Do something else, Little John.

Norton Air Force Base. Bill Crabtree is a kind of a protégé of General Kent. Glenn Kent: Systems Command, Assistant Chief, Studies and Analysis. Kent is legendary himself. Toomay: "If you want a guy to think through a question, and only that one question, Glenn Kent is the man to get. Then, once he decides, he's immovable." Kent tells the story of Crabtree. There he was in this darkened room, at a briefing. Over at the Study Facility. Some general talking. Every once in a while Kent hears this little voice in the back. "General, it doesn't quite work like that." "General, those radars don't operate in that frequency." He looks around. Can't see who it is. Voice is still there. "General, there's a better alternative to that approach." Finally the briefing comes to an end. Kent goes looking for the voice. But there's nobody back there except this little lieutenant. Turns out it's Crabtree. So Crabtree gets major below the zone. HE'S going somewhere!

Michael didn't notice when the rescue squad ambulance returned. The snow had almost stopped falling. The world outside was all a lamplit black and white.

Norton Air Force Base lies baked in the summer sun. Hepfer is working on the bus which drops off three RVs on separate targets. Zeiberg and Toomay are making sure both single and multiple RVs get through. Crabtree is working on multiple RVs that not only head for separate targets but zigzag on the way. Multiple Independently Targetable Reentry Vehicles — MIRVs — the concept Latter told Brown would work in spite of Mc-Namara's skepticism, are an accepted part of the world of nuclear war. And although the Russians haven't built them yet, they are already an accepted part of the threat.

Coughs, rapping on the table, a young briefer, blue suit crisp as a new fence. Gentlemen: We have reliable information that we have designed, perfected and deployed Multiple Independently Targetable Reentry Vehicles. This technology could endanger our silo-based missile force. Gentlemen . . .

There are no voices in the back. All agree.

Albert Latter: "Without MIRV there is no threat to the U.S. land-based missiles." With MIRV, even in the country's own hands, all those Minuteman missiles hiding in stationary silos all over the place — some destined to be carrying MIRVs themselves — are in trouble.

Norton Air Force Base. How much skin does an RV lose on its way back through the atmosphere, and where does it lose it? How about this: Put radioactive isotopes on the surface, some telemetry gear inside, and watch the isotopes ablate off. Great, let's do it! Stick it on an Athena at Green River and shoot it down to White Sands and watch it — WHOOPS!

Damn. Where did it go?

9

I

"WE had some peculiar conditions under which the reentry vehicle had high erosion rates," Robert Duffy said. "And we used radioactive isotopes embedded in the reentry shield. As the radioisotope particles were removed you could tell how the nose tip was recedin' and the rest of the skin of the reentry vehicle. We fired that one and it didn't come down."

Duffy paused to light his pipe. His cheerful, almost droll face would have been constantly shrouded by smoke, except that the unobtrusive air conditioners in his office swept his aromatic clouds away. Duffy's entire building, a huge, black and white, glassy new structure near the Massachusetts Institute of Technology, was one of the cleanest buildings in the world. In rooms whose entire air content was constantly circulated through acres of filters, men and women in surgeons' clothing created and tested instruments more sensitive and balanced than the human inner ear. Robert Duffy was president of a company called Charles Stark Draper Laboratory. Draper designed the relatively small and precious box that was the center of what was known as the guidance system. This piece of equipment was smaller than any one of the ten RVs the MX would carry, but more important. Probably at Duffy's good humored instigation, people at Draper liked to compare the dimensions of the guidance system to those of a small keg of beer.

Outside the windows of Duffy's office steam from vents on the roofs of other buildings blew north in a bitter wind. Michael's face was still raw from walking in the wind. All the radio stations in Boston were calling for snow. In the clear air outside everything was ice.

On a blackboard behind Duffy were scrawled notes and messages. Only two had meaning to Michael: "Call Dr. Atkins." "Call Jack Welch." In 1971 Duffy had come here almost directly from Norton Air Force Base. There he had been the general in charge of ABRES. One source had told Michael that one of the reasons Duffy retired as only a brigadier general was the loss of a missile.

Duffy looked at Michael deadpan, the humor showing only in his eyes.

"It was not exactly a triumph," he said.

Michael had a transcript with him of Seymour Zeiberg's remarks on the missing missile. "The Athena was a funny missile," Zeiberg had said. "It had no operational significance but it would allow you to test the reentry vehicle. Early on in its development one went astray and landed close to Durango, Colorado. Then, in Duffy's tenure, we lost that other one."

Duffy released a cloud of smoke. The smoke was whisked away, leaving the room as clear as the arctic mass outside. He continued.

"The Army guy who runs the range was in the control center at the range with me," Duffy said. "He said, 'Hey, we're not getting anything in the impact area.'

"I said, 'Well, call in the two big radars.'

"And the guy there said, 'The last time we saw it it was at about three hundred thousand feet and had a heading of about one-seven-zero degrees,' which was just a little west of El Paso. And its velocity was fourteen thousand, twenty thousand feet per second, whatever."

Duffy puffed meditatively, glancing at Michael as if to make sure the story was earning appropriate respect.

"So we got the velocity at that altitude, looked at the maps, made a quick plot of how far it was able to go. I had to call, quickly, to the command post of the Air Force to say we have an impact in a foreign country."

Smoke rose and vanished.

"The guy says, 'Shit.' "

II

On Duffy's desk was a small model of a missile and a softball. Outside the steam from the tops of buildings raced past in the wind. It had been a chill day indeed back when Duffy had lost that Athena.

White Sands, New Mexico. Three A.M. The rules for catastrophe require certain calls be made in appropriate order: Now the telephone is ringing at United States Air Force Systems Command in Washington, where a general named Ferguson is next up the chain. It is 5 A.M. in Washington.

"Good morning, sir!"

"Hello. This is General Duffy."

"This is General Ferguson's duty officer, sir!"

"I have an urgent message."

"Yes sir!"

"I have a report to make to General Ferguson."

"I'm sorry, General Ferguson is not available right now, sir!"

General Duffy is a pleasant man. His worst enemies would be forced to describe him as genial. He smokes a lazy pipe. He likes to call people by their first names. He goes to every sporting event in which his troops play. "They're my kids." They give him souvenir softballs. He is comfortable around the middle. He did not concern himself overly with protocol. He says to the duty officer:

"We have a broken arrow incident."

"I'm sorry, sir!" says the duty officer.

On the telephone line across the country explicit descriptions of such an incident are forbidden. The duty officer has apparently missed the significance of Duffy's words. Duffy can only give him the basic coded report and hang up.

In his office at Draper Labs Duffy picked up the telephone. "Kathy, ol' darlin,' can you bring up the latest annual report? Wonderful." At Draper people called him Duff. Everybody there wore official badges bearing full-color photographs of themselves that made most of them look like prisoners; on Duffy's badge his face was partially obscured by his pipe. Duffy leaned back thoughtfully, making a vanishing cloud. Outside the window a jet airliner bound for Logan Airport slid past the window in a silent trajectory.

White Sands. Five A.M. Duffy has conversed at some length on the telephone. The State Department is now involved. An aide nudges him.

"General Ferguson on the line."

The general has at least been informed.

"Good morning, General Ferguson."

"What the hell's going on down there, Duff?" Ferguson does not sound sleepy. "I've been trying to reach you for an hour."

"I'm sorry, sir."

"Where's that missile?"

"Well, I'll tell you." Duffy's pipe is cold. "We don't know."

Seven A.M. is a difficult time of day in summer in Washington. Just the promise of the pending heat of the day opens operations with a threat of its own.

"Come on, Duff. Get off that horseshit. Where is it supposed to be?"

"I'll give you exactly what I know. The last thing we knew it was at about three hundred thousand feet, it had a velocity of fifteen thousand feet per second, and its heading was one-seven-zero degrees. My guess is it's in Chihuahua."

Some of that humid Washington heat steams in over the line.

"General Duffy," Ferguson says. "Don't get smart with me."

Everyone at White Sands is familiar with the characteristics of a ballistic missile. It has a limited amount of fuel. It uses it up quickly. The rest of its journey is spent as a bullet, fighting gravity and friction with momentum. And unless orbit is reached, which is most unlikely with an Athena, the missile inevitably returns to earth. Shortly after the disappearance of the Athena there may have been a few moments of strained silence in the control room in between telephone calls, when Duffy and the others there might surreptitiously have listened for the distant whump of an RV hitting the ground, but by now the end of the flight and the beginning of some new drama has become certain: Somewhere in the unrespon-

sive silence of that early morning the business end of that
Athena missile has abruptly come into contact with
Mexico.

Norton Air Force Base. Early morning sun. Clear air.
Palm trees. Shadows that stretch for miles. The thunder
of departing C-141 transports. The smell of burned kero-
sene. Men reporting for work at the Ballistic Missile Division, at Aerospace. John Toomay. Seymour Zeiberg. Bill
Crabtree. John Hepfer. Good morning! Have you heard
about . . . ? Have you heard? Have you? Well, where's
it gone? Mexico.

But the change of the decade is the beginning of the
end of a good period of life. Soon John Toomay will be
promoted to general, and will be sent to the Rome Air
Development Center. Crabtree will make major below
the zone and go to Studies and Analysis at Systems Command to work on the development of the Air Force's
A-10 aircraft. John Hepfer will be transferred back to
Washington to be the Minuteman representative at Systems Command at Andrews Air Force Base. Shortly
General Duffy himself will be reassigned to Washington,
and, not long after that, will retire from the Air Force
and move to Draper Labs. In 1971 Aerospace will decide
to close up its operations at Tippecanoe and Waterman
in San Bernardino and move back to Los Angeles, and
at that point Seymour Zeiberg and others will start re-
assessing their career goals. Zeiberg will listen carefully
to the possibilities offered by an analysis group being
formed called RDA, and when the Aerospace move is
made in 1972 he will go to work there for Albert Latter.
Pitrozello's Del Fino de Oro will find itself abandoned
among the bars and pizza joints that claim the Norton
weenies, and will close down. ABRES itself will decline

in importance, as decoys make way for MIRVs. Anyway, what other achievements could you hope for, after all these years of working to hide warheads, when you can't even find your own?

White Sands. Ten A.M. A harsh morning in New Mexico. After various international consultations, it has become apparent that the missile has not hit a populated area. In fact nobody at all saw it streak down from the morning sky and hit the desert. Missiles do not go straight up and come straight down, PLOP, although that's always what it looks like in diagrams. They come in at a shallow angle, so shallow that one suggested missile defense consists of small forests of steel stakes stuck in the ground uprange of the protected silo; the stakes, only a few meters tall, would batter an incoming RV so it would either crash or explode out of range of its target. The Athena's RV has not made a crater in Mexico. It has plowed a furrow, in which it has sown shreds of radioactive metal.

At Draper Labs Robert Duffy put his pipe on a dish and leaned back in his chair, folding his hands in contentment. He gazed happily out the window.

"We got the whole goddamn Mexican Army out looking for it for a couple of days," he said. "Then I gave up on that and called the Air Force Weapons Laboratory and said, 'You guys are alleged to have a detection capability for this sort of thing. Can you lend us your airplane?' We found it in a matter of a day after we got that airplane down there."

The Athena warhead had hit the ground not too far from the substantial city of Durango, Mexico.

III

Michael strode through the hallways of the Draper Laboratories with Robert Duffy. Duffy spread enthusiasm as if he was operating a firehose. Hiya-John-how're-you-doin'. They glanced in windows at people in lab coats who looked up and waved. Duffy showed him the beer keg. The description was accurate. It held a set of three gyroscopes and three accelerometers in a ball that floated in liquid. Michael had already filled his notebooks with metaphors for the accuracy this new device provided. It was about as good as that required by a baseball pitcher who stood on the mound in Fenway Park in Boston and prepared to throw a strike to a batter at the plate in New York City's Yankee Stadium. It was as good as kicking a field goal in the Rose Bowl from somewhere around San Diego.

"We just design the systems," Duffy said as they romped through the corridors. "What we get our high out of is doing it the first time and then coaching someone else to produce it."

They stopped in a room where massive tables rested on concrete.

"We built this building around the test stands," Duffy said. "The stands go down into the earth. They can be accurately aligned with the earth's axis."

They strode through the halls. A cockpit of a 747. Fault tolerant computers. "We've got fifteen or twenty Air Force officers here working on their graduate degrees." F-8 cockpit. Another test stand. "There's an MX-class gyro on that stand." "Hiya-Fred-how're-you-doin'?" A class 100 clean room, where air flowed from the ceiling to floor carrying away invisible particles as if they were germs of sin.

"We get pickets here every Monday and Friday. I talk to them. They're good people. We've never had anybody say they were not going to work here because we contribute to weapons. To be honest a few people probably have left for that reason. But I respect those folks."

The beer keg, open. Inside, a beryllium ball holding gyros. Wires on the inside of the shell of the keg. Duffy explained the system. Inertial guidance was nothing new. People had been working on it for decades. Duffy had worked on it since 1953. So had John Hepfer. What was special about the MX system was the new way it was designed to avoid problems that led to inaccuracy: friction and heat. The new unit, called AIRS — Advanced Inertial Reference System — was a ball that floated inside another ball full of liquid that was not beer. Inside that forgiving embrace, the gyros and accelerometers were warm, comfortable, tremendously stable, and terribly accurate. The system, Duffy said, treated its recording instruments with the tenderness of a mother carrying her unborn child: AIRS was an embryo, a beryllium baby, nestled in a womb.

Michael wrote down the metaphor.

They strode back to Duffy's office — Hiya-Jim-whatcha-been-doin'-lately? Michael was subdued. He watched Duffy pump up his vessel of talent at each point of human contact. His eyes were blank. He was seeing again the birth of his son, and the afterbirth: the marvelous placenta that had cared for his child so long. The placenta, with its gorged veins. Parts of the AIRS keg had indeed looked like the placenta, with their linings of red, blue, yellow, and green electrical wires. As Duffy passed and spoke to a succession of cheerful men and women, Michael stared at their faces with somber eyes.

Duffy was chuckling at something. The end of the Athena episode. There in the desert near Durango the remains of the RV lay, in a new-cut trench. The RV was fragments. It was dust. "The radiation down there was about as much as a half-dozen radium watches," Duffy said. "But the State Department made an arrangement with Mexico that they'd take a trainload of the dirt around this place and deposit it in White Sands National Monument, which was done. A trainload!"

Michael looked at the small, cheerful, buoyant man walking beside him, and grinned. Athena had returned. In his transcripts of Zeiberg's remarks Michael possessed a final conclusion.

"There is a third Durango in the world," Zeiberg had said. "It's in Spain. People said 'No, you can't do that!' Well, one guy sat down and did some calculations. If the payload was small enough, Athena could put something in orbit, and if the guidance system went awry it could conceivably come down in Durango, Spain. So we showed that to Duffy one day." Zeiberg had smiled, with great amusement, with great affection. "Duffy got historical over that."

IV

The two men were back in Duffy's office. Duffy worked on his pipe. Michael asked:

"How do people like those at Norton develop their feeling of responsibility to the — world?"

Duffy contemplated briefly. The pipe came to life.

"In all honesty," he said, "I think the young people who came to us first were fascinated by the technical content of the work. That wrapped them up. In all honesty I'm not sure that at that early stage of their lives they had formed hard opinions. But as they got exposed more and more over the years they began to perceive that threat. If you're exposed to information — privileged information which the general public doesn't get — you can't help but get a little concerned. And once that concern begins to build, you look, and most of us came to the conclusion that safety was in creating a counter-threat that was awful. I use that term advisedly. I really mean awful."

Michael scribbled in his notebook. Duffy's voice went on.

"The group that was out there at that time all saw a real danger and they were willing to put in the effort because it was clear to that group and to their leadership that they had the welfare of the country in their hands." Duffy paused. Michael wrote. Duffy continued. "You can make that even bigger but most of the people who are involved in it don't think that big. You can see it as the welfare of Western Man."

Michael stopped writing. His little tape recorder still ran. Michael looked up at Duffy. He stared at Duffy. The man's face had changed. His friendly eyes were cold. His buoyant cheeks seemed to have gone flat. His lips were tense, even as he spoke. When he paused they shut in a hard line. Michael put his book on his lap and watched Duffy.

"— or even the welfare of the world," Duffy said. "Because I think all of us felt that we didn't think that the institutions that had been developed over the centuries that made for Western civilization should be washed out by people who had a different philosophy of life than we had."

Michael listened, his own face going cool and without expression.

"I think there is that difference between the West and the East," Duffy said. He leaned forward. "Our people lose sight of the fact that those missiles were rattled by the Russians first — We've always been in a reactive mode.

"I think we did some stupid things in Vietnam. I happen to have spent a year in Vietnam. I think the stupid things were done with the best of intentions in the world. In the time that we got ourselves diverted from the real threat, our potential adversary just kept right on going with his strategic capabilities, and we stopped cold.

"And what a shame. While we stopped cold he built up and now he's a formidable foe."

For a moment his voice softened. He seemed back in character, the cheerful, kind, happy man. "I still believe in the strategic disarmament and I think that that could be the salvation of civilization as we know it," he said.

Then again his lips became thin and white. "But I can't see us negotiating by going with our hands open to those guys and sayin' We Love You and here's the other cheek. 'Cause they don't understand that, at least from what I've been able to understand about the way they've reacted.

"Think about the line of countries that are Russian or Russian dominated these days, versus what they were before World War II. You can't close your eyes to that. Now you find them in Africa, their influence certainly in this hemisphere. Stronger than it's ever been. It's a worrisome thing. And I'm sure people in Russia are worried about what they see as aggressive activities on our part. Unless we can cool that thing off I think we've got troubles."

His face was rigid. He spoke carefully.

"One way to cool it off in my estimation is to build a strategic arsenal which is —" he paused to pour meaning once again into the word, "— awful. I . . . mean . . . AWFUL."

Michael stared at Duffy with his large eyes. He said:

"I understand — you mentioned that sometimes you've been picketed here. Isn't that partly because AIRS — your guidance system — permits the United States to make a first strike? To hit them first? And those people think that's immoral? What is your reaction to that argument?"

Duffy considered briefly. Then he said, hard and cold again, perfectly determined: "I think we should provide the leadership of the country with all the options that it needs to preserve our way of life."

Duffy went down to the lobby with Michael. Duffy chatted ebulliently with a couple of men in the elevator. Michael was silent. At the bottom Michael said:

"Thank you for your time."

"Enjoyed having you." Duffy grinned.

"I hope that my work will be — worth it," Michael said vaguely. He watched Duffy, this small, buoyant man: He didn't have his pipe with him, but there it was on his pass photo. "I hope it expresses my — interest in this subject. I hope to write something that is . . ." He looked at the man behind the guidance of the bombs. "That is — ACCURATE."

Duffy smiled. "This subject can become absorbing," he said.

Michael walked down the sidewalk towards the Boston subway. In the bitter wind he seemed to stagger. He was looking at the ground. On the concrete landing of the stairway underground someone had recently spray-painted, with a stencil, the simple white image of a mushroom cloud. Beneath it was a one-word question: "TODAY?"

TWO

The Hybrid, Garage-mobile, Ground-thumping, Spurred, Hardpoint Zipperditch

10

STRATEGIC NUCLEAR BOMBERS

United States:	*376*
Soviet Union:	*202*

I

IN the early light of 7 A.M. in Montana, Michael cele-
brated the fourth birthday of his daughter with the
sleepy girl, his wife, and his son. They had candles on
doughnuts. It had rained in the night. As they left their
house the spring fields lay drowsing with latent power,
under a haze of moisture. The evergreens, hung with
raindrops, shimmered. The shadows in the mountains
were still filled with snow. The depth reports were all
in: there was enough water stored there now for the
whole season's irrigation. The farmers were safe.

Michael caught an early flight for Washington. He
waved to his family from the windows of the plane. He
seldom talked to his wife about nuclear war. She lay
awake. Changing planes in Detroit, Michael bought a
copy of the newspaper. President Ronald Reagan had
addressed the graduating class at West Point. In the

speech he had attacked the Soviet Union as a "nation of prisoners."

Michael tore the transcript of the speech from the newspaper and threw the rest away. He sat down to wait for the plane, reading.

In a little sidebar below the West Point speech the newspaper reported that Defense Secretary Caspar Weinberger had told Air Force graduates in Colorado that the United States "must realign our sights to the new geography of conflict while not losing our ability to deter and to defend in more traditional theaters. . . . As a result of these changes many of you will spend a considerable portion of your professional careers thinking about, and even serving in, parts of the world that once seemed remote and irrelevant to most Americans."

He turned to the transcript of Reagan's speech.

"Those shrill voices that would have us believe the defenders of our nation are somehow the enemies of peace are as false as they are shrill," the President had said. Near the end, just before a salute to West Point using Douglas MacArthur's words, Reagan had added: "Today you are (a) chain, holding back an evil force that would extinguish the light we've been tending for 6,000 years."

Michael got up and walked around the airport until he was called to board the plane. In the air between Detroit and Washington, as the jet flew between towering clouds of spring thunderstorms, Michael wrote a long letter to his wife, expressing his love.

The following morning he was in the Pentagon.

II

He was early. His appointment with Marvin Atkins was for almost forty-five minutes later. He did not get entirely lost. He found Atkins's office and then backtracked, exploring the building. He found himself in a hallway of paintings. A plaque near faded slides of fireworks named the hall the Bicentennial Corridor. "Dedicated May 7, 1976."

The display consisted of a series of paintings, hung behind a red, white and blue rope. Before each painting stood a brass plaque with a description. Michael started at the far left and looked at every one. At each one he wrote remarks.

"New Providence Landing, March 3, 1776." Palm trees, green grass, sunshine, blue sky, one ship, firing cotton.

"Trenton, December 26, 1776." Snow, houses, stone walls, men firing long rifles.

"Yorktown, October 14, 1781. . . . An event that virtually assured American Independence." Michael stood a long time in front of the painting. Green fields, puffy clouds, a fence, a few soldiers, a smoking cannon. A wounded man, looking into the left distance, feeling inside his shirt absently for the source of a gush of blood, with eyes of intense contemplation. Michael stared at those eyes.

"Burning of the Frigate Philadelphia, February 16, 1803."

"Army Pathfinders of the West, 1804–1807."

"Battle of Lake Erie, September 10, 1813 . . . 'We have met the enemy and they are ours.' "

Behind Michael the lifeblood of the Pentagon flowed: little yellow carts, officers in immaculate uniforms, well-dressed civilians. The feet sounded loudly on the linoleum floor.

"The Battle of North Point and the Defense of Baltimore."

"The Battle of Chapultepec and the Capture of Mexico

City, 1847." White and gray smoke, adobe buildings, dark blue uniforms.

Behind Michael someone was talking angrily: ". . . so it fell through the crack. Major Collins was supposed to handle it. He didn't . . ." Michael added the remarks to his book.

"Missionary Ridge. . . . The Soldiers' Battle. . . . Spontaneous action of rank-and-file soldiers." Clubbing, stabbing men, a hillside cluttered with war. Broken trees. A Confederate battle flag. A Union battle flag. A wounded soldier sitting with his back to a stump, his mouth open, gazing with an expression of dazed thirst toward the right, toward the next painting.

"The Battle of Mobile Bay, August 5, 1864 . . . 'Damn the torpedoes, full speed ahead.'" Fire, smoke, iron boats, tufts of water growing from the sea.

"Buffalo Soldiers in the Ute Campaign. . . . Trouble broke out at the White River Agency . . ."

". . . worthless junk," somebody said behind Michael. "The software is one of the big items of cost . . ." Distracted momentarily, Michael looked back at the painting, which focused on a black soldier carrying water through a rain of arrows. The soldier was firing a revolver. "Trouble broke out at the White River Agency in Colorado," the plaque continued, "as a result of Ute resentment over Agent N. C. Meeker's efforts to educate and convert the Indians to farmers . . ."

"Battle of Manila Bay, May 1, 1898 . . . 'You may fire when ready, Gridley.'" Blue, blue water, red and white ships.

"The assault on San Juan Heights, . . . the high point of the land fighting in America's brief war with Spain." Ultra green grass, ultra blue sky, a black soldier dying in the grass with blood running from his mouth.

"Panama Canal."

"World War I convoys." Gray ships, gray sky, hard blue sea.

"Stopping the German at Belleau Wood, June, 1918. . . . 'Retreat, hell. We just got here.' " Brown-green hills, men in khaki, gas masks. Early tanks firing. A machine gun emplacement, occupied only by an empty helmet.

"St. Mihiel Offensive, September 12–20, 1918."

"Pearl Harbor, December 7, 1941. . . . But Pearl Harbor served also to unite the American people and inspired a grim determination . . ."

"Doolittle raid on Japan, April 18, 1942."

"Battle of Midway, June 4–6, 1942." Aircraft carriers seen from the air. Three United States aircraft.

"The Battle for Guadalcanal. August 1942–February 1943." Three men in a ditch, a green palm jungle, a great heave of water. One man falling, one firing a gun with a huge orange flame.

"Big Week Bombardment of Germany. February 20–25, 1944." A bomber named Fancy Nancy.

"Normandy Landings, June 6, 1944." Green hills, blue water, landing craft, a helmeted lieutenant waving his arm.

"Trinity. The First Atomic Explosion . . . heralded a new era in the world environment . . ." A column of red and gray cloud, puffs of yellow, three spiky Yucca plants, in the foreground, looking like explosions themselves on the desert.

"The Berlin Airlift, June 26, 1948–September 30, 1949."

Loud footsteps in the corridor. A laugh. A remark. "We have met the enemy and we have lost." Shared laughter.

"The Inchon Landing, September 15, 1950." Gray-green land, gray smoke, fire, men running in cracks in the ground.

"First Jet Air War, November 8, 1950–July 27, 1953."

"The Siege of Khe Sanh and Operation Pegasus, January 20–April 12, 1968." This last was a deep, somber painting, all in blue-green: a standing soldier, an aircraft dropping parachutes that looked like bubbles, a twin-rotor helicopter with a net, a sandbagged bunker.

The hall ended. Michael stopped and walked slowly back down the single rank of American wars. Someone coming the other way observed, ". . . they were, playing backgammon in the computer room at lunchtime . . ."

Michael stopped again to look at the man who was feeling so thoughtfully for the hole in his chest at Yorktown, the battle that had made the United States. The painting was at almost the opposite end of the corridor from the explosion at Alamogordo. He stared hard at the man's eyes, but found no words to write in his notebook.

III

Atkins's office was on the third floor. It was a big room. Michael and Atkins sat at one end of a long table. At the other end were a viewgraph projector and a screen. In a corner of the room was a filing cabinet with a combination dial and a large sign: LOCKED. Above the cabinet was a dartboard with segments labeled "I'll think about it," "Maybe," and "Go ask Pete." Above that: "MARV ATKINS DECIDES."

Atkins was a lean, ruddy-faced man with a dark beard. He smoked a long, brown cigarette. His beard had thin patches in it, and the color in his face was bright, as if windblown. He looked frostbitten. He looked fresh from the Antarctic. He sat calmly, watching Michael with a speculative and amused look in his eyes, as if months of wind and solitude had sanded him down to a raw efficiency. But of course it had not been storms or loneliness; it had been years in the Pentagon.

He spoke in a cheerful baritone.

"At that hearing we attended," Atkins said, "it galled me that Seiberling and Santini both kept talking about unbiased evaluations over and over again. I like to think that the Office of the Secretary of Defense is unbiased."

Atkins laughed. The laugh was hearty. It brought even more color to his cheeks. It ended abruptly. He went on: "We have no particular axes to grind except that we think national defense is a good thing. Beyond that, exactly how it's done doesn't affect my pay."

Atkins's eyebrows rose from the outside, like wings, making his face suddenly severe.

"Weinberger is going to be very heavily involved in this review of MX basing," Atkins said, "because it's the biggest single thing he has to decide this year. The administration would like to make its final decision around the first of July. That's not very long. We got started several weeks ago, pulling out all the old files and blowing the dust off. There's quite a job to do in putting all the information in parallel format, updating things, and so forth and so on."

Atkins drew slowly on his brown cigarette, his eyes never leaving Michael's face.

"Certainly," he said, "Weinberger will seriously consider the public reactions to the environmental impact statement that's on the street now. He wants to absorb all that. The comments will probably be somewhat aggregated, like 'We got sixty-seven postcards that just said Go to Hell,' or something like that." Atkins laughed.

He stopped laughing. His eyebrows rose. His scrutiny of Michael remained unbroken.

Michael asked:

"Isn't it difficult for people in this atmosphere of uncertainty?"

"Of course," Atkins said. "It is particularly difficult for people like Bill Crabtree out in San Bernardino. Back here, when you read the Washington *Post* over the breakfast table every morning you get used to it. But hard-nuts engineers who are working twelve hours a day tend to lose patience." He laughed again. Michael chuckled with him. The laugh ended abruptly. Michael returned Atkins's steady gaze. Atkins went on.

"The system has been under some attack from the hawks in Congress. Some went so far as to accuse the previous administration of deliberately announcing a system that was so bad and so expensive and so un-workable that it would eventually collapse of its own weight. Those hawks really ought to pay close attention to the viewpoint of the blue-suit Air Force on this kind of thing. The blue-suiters support it to a man.

"I believe very strongly that there will be a lot more public support for the system the more they understand about it. It has worried me a lot in the past that perhaps we suppress too much intelligence data, that the national security would be enhanced if the public knew more about what we know about the Soviets. We release all kinds of information about new defense programs such as the MX, but we still say very little in public about what we know about Soviet activities. It worries me that our inhibitions about doing this do keep the public from forming what I believe would be an appropriate perception of what the Soviets are doing."

Michael asked:

"As I understand it the reason for the MX system

is the perception that Minuteman will become vulnerable." Atkins nodded. Michael went on. "Do you have a sense of the beginning of that vulnerability?"

Atkins tapped out another cigarette.

"Quite early people had the idea that eventually missiles might become so accurate that you would be able to attack silos successfully in a one-on-one attack," he said. "As a result, there was a big activity in the late 'sixties which ended in nineteen seventy with a big improvement program called the Hard Rock Silo program. That was the first real outgrowth of the study that was called Strat X. But then they were doing some field tests, and the idea got unsold, because the thinking shifted around to the idea that there were modest improvements that could be made in the existing silos that would make them survivable for another decade. That probably no silo would be survivable for more than a decade, so there was no sense in building a whole new set. In retrospect that was absolutely right."

"That was when the basing modes began to pop out?"

Atkins put down his long cigarette.

"A lot of those popped out of Strat X," he said.

"Were you involved in Strat X?"

"I was only vaguely associated with it. I think I came to one meeting. I was a junior guy with a defense contractor at that time, and I was hauled in here by some friend of mine who was on a panel, worked on it for a couple of days, and I didn't really know what the hell was going on." Atkins laughed, loud and hearty.

"Who were you with?" Michael asked.

Atkins stopped laughing. "I was with Avco Corporation," Atkins said. "I was an ABRES contractor." He laughed again. His winged eyebrows rose. "That's part of the revolving door. I guess the highest-level job I had at that time in my life in the program sense was running a five-million-dollar-a-year contract. It was not at all connected with the stuff that Toomay and Zeiberg were doing. The project I had was on hardening of reentry vehicles against nuclear effects. My background is nuclear effects."

"Did you work for Lloyd Wilson at all?"

Atkins laughed.

"Well, he was here in Zeiberg's job, so in a way we all did. Lloyd Wilson was quite a character, a real character! Ended up shot and killed."

Atkins's laugh snapped off into a momentary silence. He watched Michael. His windblown face was still. Atkins didn't mind the silence. He spent it watching Michael.

"When —" Michael said. "What is the point at which a person takes a personal stand on nuclear weapons? Is there a point at which one decides that this is an important thing to do?"

Atkins's eyes seemed to cloud. The red in his cheeks was like a stain on cold flesh. His eyebrows rose a millimeter. He gave a slight sigh.

"Well, that's very interesting," he said. His voice sounded bored. "I'm sure each individual might give a different answer to that." He paused. Again he gazed steadily at Michael. He said, blandly: "That's a very interesting question. I don't know any all-purpose answer to that. I think it's vague."

Michael got up to leave the office. Atkins shook hands and laughed. He raised his eyebrows, and color pinched into his cheeks. He said goodbye.

Out in the Pentagon, Michael went looking again for the corridor with the paintings of the United States' wars. In the time he had, he couldn't find it.

11

I

MICHAEL drove down the east side of Chesapeake Bay in a rented automobile. He drove too fast, as if he had to get somewhere. The fields here were already green. The deciduous forests, fully leafed, looked as deep as jungle. Sunlight and wind. Sunlight and sails. Sunlight and new growth. Sunlight and children eating ice-cream cones in the park across the street from the soda fountain at the old town of Oxford, on the shore.

Waiting for a ferry, Michael watched a group of sailboats tacking upwind, mocked by a windsurfer, whose toppling sail was disconcerting among all those sails which must not fall. He watched the boats turn at an invisible buoy and then race back, seeming to flow in the current with the windblown grey-blue water.

Michael found a room at an oddly incomplete resort that looked as if construction had been interrupted several years before. Building sites, already bulldozed, were overgrown. The paved road was cracking. The enthusiasm had been bleached out of the signs. The hotel, at least, was finished, and he put his piles of notes and documents down in a room that looked out across an expanse of rough grass to the water. Nearby, on the edge of the view, was one of those bladed, unbuilt sites. It didn't look as if someone had hoped to build something there; the emptiness had substance to it, as if there had once been a home there and now it was gone.

He sat at a tiny desk, spread out his notes, and tried to build a recollection of Strat X.

Washington, D.C. 1967. Strat X is a study generated by Robert S. McNamara. It is designed to outline the future of strategic missile systems. Strat X is like so many defense studies created to knead a tough question: an assembly of smart people yanked out of less exciting situations and suddenly charged with responsibility. It is a grand version of the kind of intense group effort that is known as a Tiger Team. It has happened year after year since that greatest Tiger Team of all, the Manhattan Project, but the sound of that solemn charge still fires the mind.

Assembled men: You understand our problem, you know it is unsolved. You are our soldiers of the intellect: Make up this new system; think up this thing, this fabrication of steel, concrete, electricity, dying metal, this utterly new idea, this venture into the unknown that will solve our most pressing dilemma. Produce the invention that will save your nation!

Strat X. A team effort. A wonderland of inspiration. Strat X draws everyone to play in its light. Sooner or later most everybody stops by the Institute for Defense Analysis building across the Stirley Highway from the Pentagon, where Strat X is housed, if only to salute; if only to brush against that excitement. Atkins drops in, in tow, his laugh a bark. Hepfer and Crabtree contribute guidance knowledge. Toomay, who thrives on all-out team efforts, and Seymour Zeiberg, who is rapidly becoming a defense insider, offer reentry information. Al Latter sprinkles wild ideas like radioactive seeds. Strat X is the place for dreaming physicists, for engineers, for soldiers, for the hard workers, for those who like to believe they stand between their nation and the abyss.

And the ideas! How do we hide our missiles? It's a wonderful question. Open-ended, rich, promising. How can it help but engross these minds?

How do we hide our missiles? Put them in trains, in boxcars mingled with loads of grain, of cattle, of refrigerated beans! Put them on separate rails, howling alone through the countryside! There goes the 6:15 missile. Dig them in below the south sides of cliffs and mountains! At the low angle of approach of the Soviet warheads, they'll never be able to hit them. Or, here's one: Anchor them in canisters offshore! On a sonar command, the canisters float gently to the surface, then BLAM! off they go. Put them on the highway in special trucks; put them on the highway in trucks that look like moving vans! Don't say RUSH . . . Or, how about a fleet of dirigibles, blimps, drifting along who knows where across the United States and out to sea! Forget roads — put them on a huge off-road vehicle — no, better yet! On a ground effect machine (GEM) that could carry a missile all over the southwestern desert! Stick them in barges on the Mississippi! Stick them down under the mountains in abandoned mines! Stick them in canisters in deep holes under columns of sand; after the bombs fall run water in the sand, and the canisters will float to the surface: BLAM! Fill a bunch of holes with boosters — first, second and third stages, all ready to go — and truck the warhead around! Hey, how about an amphibious airplane, about the size of a transport jet, that can go out to sea, land, wait around out there for a while, then fly somewhere else! Call it Sea Sitter. Or maybe just the big jet itself, cruising along with a whopping missile in it! Drop the missile out the back: BLAM! Call it Airmobile; there's something the Air Force ought to go for — it never liked these stationary missiles anyway.

Say, a trench! A tunnel under the earth — never see the missile, runs around on its own, breaks out to fire! Naw, forget making it mobile — forget MIRVs! — why don't we just build twenty thousand little tiny missiles — one warhead each — and scatter them in silos all over the nation? MIDGETMAN!

"It was fun." Jim Drake. Michael had interviewed him in Los Angeles. He had been a leader at Strat X. "Designing is always fun for me; it's a lot like solving puzzles; but this is a lot more delicious puzzle, because it has never been solved."

In his motel on the Chesapeake Bay, Michael sorted notes and tape recordings. Wind lashed trees on the bay's edge. The day was ending. Jim Drake was unusual. Jim Drake was one of the inventors of windsurfing. He wrote articles for *Surf* magazine. Marvin Atkins: "Have you ever seen Jim with his black suit on? All in black: shirt, pants, belt, everything." Atkins had laughed.

"No, I haven't," Michael had said. "What does he do that for?" Atkins had laughed again. He had said:

"Just to be flashy."

"I was thirty-seven or thirty-eight," Drake had remembered. "There was still a feeling of genuine exhilaration with the bloom of technology. We were still motivated by the prospect of new technological ideas; the prospect of going off and helping in a great project that might help deterrence and prevent the failure of a great strategic system was very appealing."

Washington, 1967. Strat X. A year in Washington. Ideas! But then, HA! You have to turn the ideas over to the Red Team. The Red Team is always there, like conscience. The Red Team is the in-house Soviet Threat. Its job is to figure out where your weapon idea is vulnerable. Sometimes a Red Team is specifically created for one task, like finding the weaknesses in Strat X ideas, and sometimes various different arms of defense research acted as Red Teams for each other. ABRES is the Red Team for the ballistic missile defense people and vice versa: Ha! I've got a new radar — I've got a new decoy. I've got a BETTER new decoy — but check out THIS radar. OK, put your radar down at White Sands, and we'll shoot an Athena at it; THEN let's see if you can sift out the decoys — You sure you don't want me to put my radar in Spain?

Camaraderie and good times. So, says the Red Team in the IDA building; you think you've got your missiles hidden, do you? Now how are you going to do that?

Dirigibles? OK. Barrage attack: pop 'em all! Trains? We've got some guys out there with crowbars; and what are you going to do with the public outcry when one of those things derails? Sure like your idea of putting the canister in the sand: I'll pop my little warhead anywhere near your sand and turn it all to glass. You put them down in those mines, and I'll bury them forever. I'll pluck one of your offshore canisters off the bottom and take it home. I'll shoot submarine missiles at your south side silos from just off the East Coast. I'll let you go bankrupt building thousands of boosters. I'll let you make dust beacons for my satellite photographers with your off-road-ground-effect machines. Sea Sitter? Midgetman? Come *on.*

Albert Latter, his quick, sharp face visible in the rooms or in the minds of the men here he knows, buries two main ideas in the fertile earth of Strat X. One is to store a missile on a truck in a kind of garage placed at the hub of a bunch of roads to shelters; when the warning comes the truck can dash to a shelter selected at random and pop the missile in it. The odds of the Soviets hitting the right shelter are slim. The other idea is even simpler: Put the missiles in pools.

It is a physicist's dream, a simple idea based on a couple of fundamental principles: water protects something at least as well as putting it underground, but is a lot easier to get out of; and water has no shear force: It won't pull something apart by itself the way shifting earth might. If the warhead doesn't hit the pool perfectly, the water will protect the missile. When the cloud has blown away, the missile will bob to the surface, and — BLAM!

Marvin Atkins: "Albert Latter is what we call a three sigma nut on the nuclear hardness of everything. He's a delightful guy, and personally one of the finest individuals I've ever known. Very good physicist. But Al has his own ideas of what ought to be done, that flavor everything. Actually the pool isn't all that bad. As far as the hardness is concerned I've always thought it was very good. But it was probably environmentally worse than most. All that water."

Strat X. Blue team builds, Red Team tears down. When both are sated Strat X becomes a collection of reports. In that last period, while the data is squeezed through the fingers of the costing experts, Jim Drake and another scientist spend their late afternoons in and behind the water-ski boat owned by one of those desk-bound experts. "All through the month of June we'd get a six-pack, some tonic water, a fifth of gin, fill the boat with gasoline, and go ski up and down on the Potomac south of National Airport." Flying on the water between banks of bureaucrats, they laugh at the delicious puzzle.

What happens to all those wonderful ideas? Dirigibles, Midgetman? Most of them have been shot down or blasted by the vigilance of the Red Team. Others have varied lives. The trench goes into the files for future use. Pools and canals remain alive only by the force of Albert Latter's determination. The idea of doing something that odd doesn't interest the Strat X director, a man named Fred Payne. Of Fred Payne, friends have only one definition of character: "He sails."

Jim Drake: "I remember Fred and I sitting down. Fred said 'I know silos are the right idea.' It was based on his judgment on intangibles — willingness of the service to sponsor the idea, availability of land, practicality — and lack of nuttiness. I remember he and I sitting down to discuss these, in particular on the canal idea, which was the cheapest, fastest, and easiest system of the bunch, so some reason for its rejection had to be manufactured on other accounts. Fred gave me a clue. He hit on the notion of strangeness. If there was a lot of it it was bad, if there was not much it was good. On that scale silos turned out to be the right thing."

Fred Payne sails away, leaving Hard Rock Silo behind as his Strat X legacy. It isn't a mobile scheme; it is protection: Drill a hole in granite, pop in a missile, cover it up, and, when the first salvo has blasted you, bore out from the inside — BLAM! And for a while, that is the favored idea.

One other concept comes out of Strat X, looks around like a groundhog in February, and gets an immediate Bronx cheer. It is a variation on a Latter theme. Move missiles from shelter to shelter, so nobody will know which shelter holds the missile. It is called MPS. It pops up into the chain of command, where it is soon spotted by the Strategic Air Command. John Toomay: "The Implacable SAC." The chief of the implacable SAC is a one-time fighter pilot ace from Tennessee, General Bruce K. Holloway, who is known as Cinc Sac. Holloway hops around in his office, laughing. Run the little missiles all over the place? Hide them in holes, then run them out again? Come on, this is the *Air Force,* not a *rabbit warren.* Holloway: "A terribly wild scheme. Just preposterous! A lot of these scientists have no appreciation of real-world logistics!"

Chesapeake: On the afternoon of the following day the sailors had come off the water and gone home by the time Michael drove back up the bay in the afternoon sun. All day, it seemed, he had driven past wide, green lawns, through arches of leaves, past shining water, in and out of the friendly shadows of trees. Finally, turning away from the traffic, he drove into the heart of Washington. The tourists had left; the week had not yet begun. He came upon the Capitol from behind, and stared with sudden affection at its wide green lawns, its trees and shadows, the sunlight on a crescent of dome. With the light almost directly behind it, the building itself seemed pale, a ghost, hidden by haze and dazzle, but the sun had caught a fountain on the edge of the grass, and against the indistinct mass of the building the little heap of tumbling water burned as bright as blood.

II

In the enormous reading room of the Library of Congress, Michael sat with a pile of books. The room rustled with the turning of pages, the ancient sound of the passing of information from mind to mind. Michael's book was a record of a congressional hearing: Dr. Seymour Zeiberg, in testimony before a closed committee hearing of the United States House of Representatives. Michael stared at the ordered pages. In them, forever engraved, Zeiberg sat in the crowded room, with his burly face, his broad body, his composure; informing the congressmen.

DR. ZEIBERG. Mr. Chairman, I would like to start off by pointing out the role of the various elements of the Triad, in that the Administration has recently reaffirmed our faith in the utility of the three modes of strategic delivery systems that we have been employing....

The Triad composition consists, as shown on this figure, of somewhat less than 500 bombers, approximately 1,000 ICBMs, and approximately 650 SLBMs. [Submarine Launched Ballistic Missiles]....

The ICBMs comprise approximately 50 percent of the delivery vehicles, but only about 35 percent of the total megatonnage. The SLBMs are more than 40 percent of the total weapons, but only about 8 percent of the megatonnage.... The force is laid down in order to maximize redundancy, that being one of the critical elements behind the Triad philosophy....

The damage achieved by the various legs of the Triad

is indicated here, according to a breakout of the different SIOP objectives. (Deleted)

SIOP (deleted) is the current SIOP. (Deleted) are well hedged in terms of distribution of the damage potential across the three legs of the Triad. You can see on the red bars that if any one of the legs does fail, we still will achieve most of the damage we plan to inflict upon the Soviets. . . .

. . . current ICBMs are considered to be our second generation — that is, Minuteman II and III, our current forces. . . . The Soviets are now in their fourth generation, getting ready to test a fifth generation.

We are considerably ahead of them technologically, however, in a relative comparison of generation by generation. Their (deleted) generation, that is, the (deleted) systems, are in a technological class with (deleted) ICBMs. . . .

The Minuteman IIs have a single warhead, approximately (deleted) with an accuracy of approximately (deleted).

The Minuteman III is MIRVed with three warheads, (deleted) and an accuracy of approximately (deleted). . . .

This graph repeats some of the data I showed on an earlier one, the SIOP lay-down. Here two colors represent first, red, the expected damage on the various target classes, while the yellow shows what we can reach; that is, the effective range of our system, and what percentage of the total target list we can physically reach and kill with our delivery systems. . . .

In the case of the soft targets; (deleted) roughly speaking we can kill most everything we reach. In the case of the (deleted).

Michael paused in his reading and looked around the room. Others read in silence. The dome, filled with light and a lofty silence, towered above him. He went back to the book. Zeiberg showed charts, glancing at the congressmen in a careful sequence.

The accuracy of Minuteman III is depicted here. The cumulative CEP, that is the circular error probable, is in the vicinity of (deleted). That means that half of the reentry vehicles would land within a circle of (deleted) radius. . . .

We are finishing the effort to get us into the position to reduce the CEP from about (deleted) to about (deleted). . . .

We will essentially (deleted) the yield of the reentry vehicles on (deleted) of the Minuteman IIIs. That is by the introduction of the MK-12 reentry vehicle. . . .

The silos have been going through an upgrade process for the last several years, where we took the hardness from (deleted) PSI up to as much as (deleted) PSI. They are not all (deleted).

MR. WHITEHURST. Can I ask you something else? I know this has been said before and I expect I have just forgotten it.

But, what kind of protection does that give you against expected megatons or kilotons we can anticipate? What distance out?

In other words, how accurate are the Soviets?

DR. ZEIBERG. . . . It is approximately (deleted) away where we start getting some pain, in that the overpressure resulting from a (deleted) burst, typical of the size of their weapons, produced in excess of (deleted) PSI. We start going into failure modes in the silos. This is significant pain. It's somewhat before that where we get the beginnings of pain. . . .

Michael closed the book. He stood up and looked around. The faces of the people in the room were calm.

He watched them for several minutes. There was no one there who looked like Seymour Zeiberg; no one with that same momentum.

He opened another book.

Dr. Seymour Zeiberg, at an open subcommittee hearing at the U.S. House of Representatives. Patience is apparent in the pace of his words. It must also have been there in his face, or in the fold of his hands across his belly.

DR. ZEIBERG. ... The feature of MX that makes it most desirable is that the exchange ratio is so high it forces the Soviets to spend their entire arsenal of ICBMs to take out 200 missiles. Their ICBMs represent 75 percent of their strategic forces. ...

This causes them to spend three-quarters of their Triad to get those 200 missiles that represent a small fraction of our Triad. So, viewing this in the post exchange context, it is a very bad move for them to try to attack the MX because they expend so much in order to get so little return and therefore the system acts as a deterrent. ...

That takes you to an extreme where it is hard to contemplate either side would lead. If they were so dedicated as to build so many as 23,000 warheads, cited earlier, if they were so dedicated to build an enormous ICBM warhead force and we had an MX which confronted them with this terribly adverse exchange ratio, they would clearly see it is the wrong approach to take. ...

MR. CARR. I guess you get into some political judgments, though, relative to the ability of their system, political and economic and military to accomplish that relative to ours. ... Let me get to the flat part of the curve. My point is politically and economically we do not like flat curves and the Soviets do not seem to mind that quite as much as we do.

DR. ZEIBERG. True; we also do not like some circumstances we are in today. We do not like Afghanistan or the adventurism on the part of the Soviets and that is another case point which indicates that unless our forces and our military posture and our foreign policy is sufficiently aggressive and robust, we will always be confronted with Soviet adventurism in a manner which undermines our position in the world and that of our allies.

The essence of an MX-like system is to address that problem.

MR. CARR. But not seriously; I say that much of what we do always tries to emulate the Soviets, and they try to do it to us, and I am glad we have not tried to emulate the Soviets by taking over Mexico.

MR. SEIBERLING. If the gentleman would yield, I wonder if Mr. Zeiberg misspoke when he said our policy, we have to be aggressive in order to deter the Soviets. The word aggressive has unfortunate connotations.

DR. ZEIBERG. I did not mean aggressive in the military sense, aggressive in defending our position in the world.

MR. SEIBERLING. I thought maybe you would want to clarify that.

For all its grandeur, there was a small-town atmosphere in the reading room. Michael got help from a slender young man in a tiny office. The man showed him indices, helped him look, and expressed anxiety that he found what he wanted. He did. There was Zeiberg again: an open hearing before the House Appropriations Committee. Zeiberg, a new administration behind him, explaining the reevaluation of the MX. Zeiberg dominated. Calm face, bulk, crewcut, thick arms, knowing eyes.

MR. EDWARDS. What I am trying to get at is whether this, in effect, is a real re-evaluation. Are we going in the right way, or are we just proceeding —

DR. ZEIBERG. I can assure you it is real. I had occasion to defend this basic concept in recent days, and I can assure you that there are serious questions in the minds of a lot of the new players. . . .

. . . there is, in a sense, on the part of many people and in the minds of many people, some very deep conviction that Multiple Protective Shelters is the right thing to do.

There are others who look at it and say it seems bizarre and there must be a better way, without having this technical background. . . .

It is a really difficult problem. There are no simple and easy solutions. As one of our advisers, in fact I think it was Dr. May . . . once said this is a question of what is the least rotten apple in a barrel of rotten apples. Anything we do will be expensive and complex. But that is not because we choose to develop expensive or complex systems, but rather the Soviets present a threat to us which is driving us out of the cheap and simple way to do business.

Many of us have the conviction that we have to offset what the Soviets did. We have to take a step that shows to them we won't sit still for being driven out of what we would choose to do and moreover something which makes their efforts all for naught.

That is what I mean by offsetting. MPS does that. It puts the Soviets in a quandary because they can no longer keep our deterrent in their pockets. They have to worry about the utility of what they have and whether or not their deterrent is at risk and sustainable in a war.

MR. EDWARDS. Is that part of the reason for your decision to proceed, even as this review process goes on?

DR. ZEIBERG. That is a personal speech.

Michael looked up at the quiet room. He got up abruptly and returned the books. He strode quickly away and out into the rushing street.

12

STRATEGIC DELIVERABLE BOMBS
AND AIR TO SURFACE MISSILES

United States: *1926*
Soviet Union: *260*

I

MICHAEL flew west, above a shrouded world. He put his forehead against the plastic window and stared out, down into layer after layer of pale blue cloud, and up into the pale blue inside lining of the sky's precious shell.

He tried to read the airline magazine. He ate half of his lunch. He stared out the window again. Finally he got out his notebook, and wrote in the back:

"Crabtree. What about Crabtree?"

On the ground at Salt Lake City he called Neil Buttimer again.

"How are you?" Buttimer said, with mild interest.

"Great. How's Bill Crabtree?"

"Working hard."

"Can I get some time with him in the next few weeks?"

"I'll work on it."

"Thanks."

Michael hung up the telephone gently.

II

On the top of one of the isolated buildings at the Thiokol plant a red light was flashing urgently. It made the little building look like an ambulance. Michael watched it as he and his escort drove across the hill toward Test Bay T18. The eye of his escort passed over the light with no concern. They drove away down the hill.

The test bed was a huge steel framework set into the side of a building, facing a stony hillside. Gantries bearing cameras faced it; it held the yellow cylinder of the first stage of the MX missile in a rigid embrace. When the test bed was being renovated recently, Michael's escort told him, a pair of golden eagles had built a nest in one of the camera gantries; the company had held up work for a year while the pair mated, hatched its eggs, and nursed its fledglings into flight. The escort, whose name was Charles Shown, liked the story. It was the first thing he said about the test.

He went on to explain further. The idea was to test the strength of the MX first stage cylinder itself, which would be called upon to sustain the considerable pressure of the burning of solid fuel for the five minutes of its fiery life. The cylinder was mounted, upside down, in the test bed, wired with extensometers to record its flexing, viewed by cameras from every angle, then pumped full of water. The pressure was raised to almost 2,000 pounds per square inch, held there for one minute, then raised incrementally until the case burst. "One moment it's there," Shown said. "And then all of a sudden it's not there. When it bursts there could be as much as a million pounds of instantaneous thrust. Last time it lifted the test bed's base plate six feet off the ground." The test was called a hydroburst.

Shown had wire-rimmed glasses, and a forehead of the high kind that is supposed to be a guarantee of a refined and genteel intelligence. He seemed to be proof of the rule. He applied those qualities with a mild cheerfulness not much different from that of Jack Hilden, Shown's boss. These cadres of specialists seemed to share group characteristics: Guidance people seemed buoyant, enthusiastic, rigorously kind; propulsion men he had met were almost gentle. Shown parked the car and walked him down a ramp to look at the prepared case.

The case stood in the test bed. It looked like the entire body of a railroad tank car upended. Michael stood before it and looked up. "That's it," he said. As if announcing the most obvious part of the scene.

"Um," said Shown.

Michael looked up at it with his large eyes wide. This piece of the ultimate weapon towered over him like a symbolic granite pillar, supporting the entire invisible edifice of the world's most violent hardware.

Michael stared up at it. It was marked in big letters at every foot; the top mark read "22." It was a pale yellow; the same color as the huge bead of the canister that Michael had seen being wound on the mandril elsewhere on Thiokol's grounds months before, but faded. Wires reached up the side and were stuck to the body of the case like electrodes on the chest of a heart patient. Around the base of it worked several men wearing white and green laboratory jackets. One man was climbing to the top of the case carrying a small cardboard box. Another was adjusting a huge mirror under the case through which a camera would look as the pressure increased. Like a man preparing to back up a truck, he moved the mirror up, then down, looked back, moved it again. Shown exchanged a few words with a safety officer in a white jacket, then stood gazing up at the case.

"Nothing impresses me more than to stand here in the test bed and see it," he said. "That's just stage one. Just half the length of the missile. It's a pretty impressive sight." Michael said nothing. He stood looking up.

A warm west wind blew across the depression in which the test bay stood; on the hillside clumps of sagebrush wiggled under its caress. A dark-haired man named Mike was laughing with several others, and collecting money. "Fifty cents," he said. It was a pressure pool. The man who accurately guessed the pressure at which the case burst would win the pot. Mike seemed just a shade too jovial, laughing at the talk about 2500 and 2700 pounds per square inch.

Shown explained: "He's the designer of the case." Someone made a bet on 2450 psi. Michael stood looking up at the first stage.

"This is the thirteenth full-sized case we've built," Shown said.

"Oh, oh," Michael said briefly. "Worried about it?"

"We're not superstitious," Shown said. "We're broadminded. We get apprehensive on every one, not just the thirteenth."

Men were leaving. Michael stood looking up at the case. With a kind of spontaneous dispersal, the men were getting out. "We've got to go now." Shown took Michael away. Over his shoulder, Michael took a final glance at the case, standing there now by itself, so upright, so strong.

Shown drove down to a small building several hundred yards away from the test bed. It was an alternate control room. Inside were two rooms separated by a door and large glass windows. Inside the first room were several folding chairs and the door to the single restroom. On that door was a sign: "MEN." Inside the second room was a control console and two black-and-white television screens on the wall. One showed the case standing by itself, and the other showed the road that approached the test bed, to watch for anyone who might inadvertently wander into range of the exploding case. Shown and Michael sat in the second room. Michael stared up at the picture of the case, which was now alone.

A loudspeaker in the room boomed into life: "All personnel remain clear of Test Bay T18. All personnel remain clear of test bay T18." It subsided, ticking. A group of about eight men came into the room; some found chairs and some stood. Shown, talking quietly, as if he was awaiting the beginning of a performance by the Salt Lake City Philharmonic, told him that these were all men who had worked on the case. It was company policy to give its workers a glimpse of the outcome of their work. Again, the loudspeakers spoke: "All personnel remain clear of Test Bay T18."

In the television picture of the approach to the bay Michael could see the sagebrush dancing in the wind, and the heat wrinkling the air above the pavement of the road. There was no other movement. The picture in the other screen was distorted; in it the case was a dark cylinder marked with a bright column of numbers. The cylinder bent and wiggled at the bottom, as if there was rising heat there, too.

The men seemed too large for the room; they were

wearing plaid shirts and blue jeans; they appeared to feel out of place. One of the men said, "This is your blood, sweat and tears." The others laughed, then abruptly were silent again.

"All personnel remain clear of Test Bay T18."

A couple of men came into the building. The door slammed shut behind them. Some of the men already inside jumped.

The loudspeaker spoke again, in a quieter voice. "We are pressuring."

The men remained silent. In a moment the loudspeaker said:

"450 psi."

One of the men said: "Made it so far, Joe." There was a shiver of amusement.

"500."

"550."

"700."

"750."

Michael could hear the fans blowing in the console computer.

"800."

"850."

"900."

One of the men turned out the lights in the room. The screens took on more prominence, but the picture of the case itself was no more clear, the column of numbers standing out in a vague gray flicker. In the other screen the air danced off the pavement, the sagebrush danced in the wind.

"1000."

"1100."

"1200."

"1300."

Earlier, Michael had been looking back in his notebook at his interview with Jack Hilden here. Hilden had talked fondly of his days in the Pentagon. Hilden had been an Air Force colonel and back in 1971 he had been stationed at the Pentagon as chief of the Missile Division, Development and Acquisition plans. He had been involved in the Advanced ICBM Development Program, which was caught and almost submerged in the backwash from the Air Force's enthusiasm for its proposed new bomber, the B-1. "It looked pretty bleak about then," Hilden had said. "There was really no program for ICBMs.

"Our concern was that the basic capability in the country could be dissipated. Our philosophy was to keep something alive in case the nation needed it." That something was a new missile, larger, more powerful, and possibly more accurate than the Minutemen that were then almost entirely spread around the country. But the Minuteman was a great success: slender, stalwart, poised; the new missile was unknown. There had to be a name.

"We had got to have something to provide some recognition," Hilden had said. "The whole purpose was to get some name recognition, to get some seed money."

"1350."

"1400."

"1450."

Shown was sitting rigidly still, his arms folded. The men who were standing seemed to shift their positions with utmost care.

"1500."

"1550."

"1600."

Shown had told Michael what happened as the pumps

began to work. The case expanded, he said, by as much as three inches, and as it expanded it groaned. It groaned, squeaked, snapped, moaned, crackled. It expressed its agony. Michael watched for signs of its distress. In the screen, except for the shimmer of the camera's distortion, there was no movement. The column of bright numbers stood firm. The only sound in the room was the speaker.

"1650."

"1700."

"1750."

Michael shifted in his chair. The sound of his movement seemed loud. He stared at the case, which was shimmering there in its stand, shimmering like someone's dream of the future.

The Pentagon, 1971. There is great debate among the abandoned soldiers in the Advanced ICBM Development Program. Gods and heroes fly in and out. Thor, Jupiter, Titan, Apollo, Hercules, even the female Athena had already been appropriated. Agamemnon doesn't have that ring, and Achilles — No. Finally a lieutenant colonel puts "MX" on the blackboard. Hmm. Some tradition there. X-1, X-15. The Atlas itself was once called MX-1593. Most developing missiles had those initials in their more formal names. But to use that as the common name, too?

What would it mean? Hilden: "It could be Minuteman Unknown or Missile Unknown. We concluded that the 'X' put in an unknown future element. It didn't necessarily mean 'Missile Experimental.' The conclusion we eventually came to was it wouldn't mean anything and it could mean anything.

"I walked downstairs to the general, and I said, 'We're going to call this thing 'MX.' "

On the television screen the case stood unbroken. The hot air danced.

"1800."

"1850."

Hilden had talked about the following years. Like hundreds of his military technology brethren, he had retired and had gone to work for a related company, Thiokol. Meanwhile, the future of the advanced ICBM had become more rosy. "So in nineteen seventy-four," Hilden had said, "Thiokol made the determination that it wanted to be an integral part of that activity. It dedicated independent R and D funds to put itself in that posture. It built a test motor and fired it, and it built a full-size case and burst it."

"1875."

"1900."

"Proof pressure."

The room remained silent. Shown whispered to Michael: "That's the holding pattern." That was the pressure the case would sustain in flight.

The loudspeaker, and the men, were silent for a long minute.

"Continuing pressurization to burst."

"2000."

"2050."

The telephone rang. The men stirred. It was for Shown. He spoke for a moment and hung up.

"2100."

Months ago, as Michael and Jack Hilden had walked back to the car, under a high, windy Utah sky, Hilden had said, the gentleness in his voice undiminished:

"Our problem today is not how to build a missile or a missile basing mode. Our problem is national will."

"2150."

"2200."

Shown was nibbling on his pen. Some of the men were moving slightly; Michael could hear their feet shuffle on the floor.

"2250."

"2300."

Except for the wobbling at the bottom of the picture, the case in the screen remained perfectly still.

"2350."

"Made it past the minimum," Shown said.

"2400."

As the loudspeaker spoke the picture in the screen dissolved; to Michael it looked as if the screen had been momentarily filled by a wave and then had disintegrated itself, showing a hazy picture of the inside of the television set, in which the upright column of numbers was retained briefly on the screen like a ghost, which swiftly faded. There was disorganized movement in the picture, and Michael finally recognized it as the ripple of water down the glass that protected the camera lens and down the sides of the test bay. The case was gone.

One of the men said: "Who won the pot?"

III

A few minutes later Shown took Michael down to the bay. The others were already there, prying around in the debris with a delicate curiosity. The area was soaked. All over the ground were bits and pieces of case, some the size of coins, some as large as briefcases. This glass fiber debris looked familiar in various ways: Some looked like chunks of a wooden canoe that had been over a falls; some looked like tangles of doll's hair, or like the hay that falls off the pickup on the way to the cows, or like broken strapping tape. It was golden, backed by black rubber.

Michael picked a piece up. He looked at it with a surprising fondness. "Can I take a souvenir?"

Shown looked at him in dismay, as if sad to have to disappoint him. "Oh, no," he said, "I don't think so. We have to examine it all, you know." Michael, embarrassed, dropped the chunk.

The case itself had burst in the middle, somewhere around the ten-foot mark. Part of it still stood in the bay, and the other half had fallen forward and now lay in a puddle, lengths of Kevlar hanging off of it like twine. Around the case, scattered among the shreds, were pieces of broken mirror.

Michael stared at it. "A perfect helical break," Shown said. "That's how you want it." Michael looked carefully at the remains, as if he wanted to go up and touch the huge round side of the case, to feel its golden hardness, but he didn't. He stood on the wet tarmac, shreds of case and slivers of mirror all around him, and looked at the broken body of the first stage of the MX missile. It was like the ruin of a column that had endured centuries of rain and had at last broken and fallen into the abandoned courtyard, to reveal its hollowness at last.

Shown, telling him of a previous test, had spoken of the research that was performed upon the remains of another case that had been shredded entirely into straw: "It's hard to work with it when you have that kind of cadaver," he had said. This was a useful cadaver. There was a lot left. Michael looked at it without rancor. He smiled at Shown. Shown smiled back. Michael looked at the bits of the first stage of the MX scattered on the ground, blown apart and scattered at random like pieces of the truth.

13

I

OCTOBER 1974. The sky over Vandenberg Air Force Base, California. High cirrus. Mid-level stratus. Light turbulence. A white world. An enormous C5 jet transport flies slowly across the sky, wing flaps out. Below it is a layer of clouds; far below that the edge of the sea. The day up here is fine. At Vandenberg, where John Hepfer, recently promoted to brigadier general, waits, tapping his fingers on the table, the sky is somber. Hepfer is a small man with wide cheeks that frequently fill with his smile. His smile only twitches now.

The white layer below the jet slides slowly past. The jet has crossed the invisible coastline and is now over the sea. Slowly the doors in the rear of the aircraft open. Trim motors whine. The pilot corrects the angle of attack for the new drag. Everything seems to move with deliberation: the plane against the clouds; the roll of clouds across the sea; the fall of a small bundle on a cord from the interior of the aircraft.

The cord snakes after the bundle. The bundle bursts. A parachute blossoms. The huge aircraft seems hardly to notice the new drag; it puts its head down slightly. Slowly at first, then swiftly, the parachute pulls on the aircraft like a winch pulling a calf. From the belly of the C5 emerges the creature, a black, white and orange tapered cylinder mounted on a sled.

The load, a missile on a gurney, an orange and white battered corpse on a wagon, falls from the C5. The aircraft leaps. New parachutes open and strain. This huge assembly, larger than a tanker truck, swings slowly toward vertical beneath the white, lifting bloom.

A puff of explosive. The cylinder slides from its bier. Loose in the sky, it falls toward the cloud. It accelerates.

Again, parachutes open. It is like a trick sky dive, where the crowd screams and sighs as the joker cuts away his life again and again and then recovers under fresh silk. Beneath three tight white sacks of air the fall of the obelisk is slowed and steadied. But still it drops toward the cloud.

"They called me the summer before," John Hepfer said. "It was one weekend. I had a trip planned. I was going to leave for a couple of days or something." Hepfer put his fingertips together. "I got a call from my boss, General Schultz. He said the guys in Washington really needed something urgently." Hepfer put his hands behind his head. His enormous smile filled his face. "They needed something spectacular to happen in the next sixty days. I guess Kissinger wanted something. For the SALT talks or something."

"Some advance to be . . . noted?" Michael asked.

Hepfer put his hands down, and rubbed one elbow. "Yeah," he said. "Some publicity."

Hepfer was sitting on the patio beside his pool overlooking the inlet at Punta Gorda Isles on the west coast of Florida. Michael had flown back southeast for the in-

terview. Now he sat with Hepfer, his tape recorder on a
table. They sat in pool chairs on a carpet of bright green
plastic grass. There was a warm breeze. The water lapped
against the sides of the little pool. The breeze rang a
single clay wind bell. Hepfer was a small man, with
prominent ears and an unremarkable mouth that opened
regularly in that great big grin.

"We'd talked about things that we could do," Hepfer
said. He put his hands behind his head and leaned back,
looking across the inlet. "We'd talked about air drop.
We'd looked at loading a missile on a truck. We looked
at firing two or three missiles simultaneously out of Vandenberg. We were so used to writing papers on that kind
of stuff we never thought anything would come out of
them, they were so ridiculous."

Hepfer put his hands down. With one finger he tapped
absently on the table.

"Dick Gingland, who was my deputy, said he'd go to
Washington." Hepfer said. "He left Saturday night on the
red-eye. General David C. Jones was then Chief of the
Air Force. Gingland briefed General Jones on Sunday.
Jones liked it." Hepfer laughed, and shook his head. "He
liked it! So Dick went to see Schlesinger on Monday, and
got approval. When Gingland called me and said they've
got approval for it, I couldn't believe it. They wanted it
done in sixty days."

Out on the concrete-lined inlet beyond the pool a
pelican cruised. Mullet jumped occasionally. The sun was
a white globe in very high clouds. Michael stared off into
the level distance, past the silent houses on the opposite
bank.

Norton Air Force Base, late summer, 1974: Good
Lord! Sixty days. OK. Send a team down to El Centro,
that Navy base down on the border. Get hold of a C5!

Get hold of some guys from the C5 program office! Get hold of Boeing. They've got some ideas. Call Thiokol. We'll need some short motors. Get some Lockheed guys down here. Off to El Centro we go. Navy does parachute testing there. Sounds like a good place to practice dropping a missile out of an airplane.

OK. Boeing's got this big sled for the missile. Slides in the belly of the C5. When you're ready, out it slides.

Forty days! Hepfer: "It was high priority. There were no drawings or anything. They had the best people they could get their hands on, and they just built it." Build it and see if it works. Slam it together and shoot pictures of it as you go along so if — God forbid — someone decides they want to do this again, you can go back to the photos and reconstruct.

In the Pentagon, John Toomay is working for a man named John Walsh in D.D.R.&E. Walsh is one of those pushing this demonstration. Toomay watches the activity out at Norton with mixed emotions. He's not sure these episodes get you anywhere. Toomay: "What did we call those things? Something out of a circus — Stunts!"

Norton. OK. Thirty days! Put the sled in the belly. Take it up and try it. Use a weight to simulate the missile. A great big, lumpy, 75,000-pound pile of metal. Looks like a bargeload of metal garbage — corrugated steel? Nuclear waste? OK, fly out over the parachute range. Slow the aircraft. Flaps out. Open the doors. Wind and desert out there. Out goes the drag chute. It catches. There goes the weight. There it goes! Chutes grab air. Whump. Weight rips itself from chutes. Whoops! Here are the chutes, fluttering down. There's the weight. Zip! There goes the weight, all alone, into the floor of the desert!

Michael: "It just went into the desert? Ka-BOOM?"

Hepfer: "Um, yeah."

OK, now we've got it working. Second test checks out. The weight is under control. Let's get some missiles. Hepfer: "We bought two or three Minuteman motors with ten seconds' worth of propellant. The guys ginned up a simple guidance system that would keep it stabilized, get it to fire straight up." Twenty days! Pack those chutes up again. Where are the pix of the last time? OK. That's how we did it. Here goes.

Vandenberg. Hepfer, in the control center, watches a television picture beamed from a chase plane. He puts his fingertips together. He chews on them for a moment. He puts his hands down and taps the desk. The doors open. The drag chute pulls the missile and its sled from the plane. The whole thing falls away. Explosives pop the sled from the missile. The sled disappears upward out of the picture. The missile falls. The new set of chutes open. The chutes catch — and hold. The missile oscillates surprisingly briefly under the chutes. Guy wires down its length keep it from spinning. It straightens out.

It's ready to fire! The whole contraption disappears toward the cloud. From the chase plane the clouds seem terribly close, with darkness showing beneath them. John Hepfer brings his fingertips together.

There is a burst of smoke. The missile, still falling, vanishes into its own white screen. It does not come out the bottom. Its hazy shape can be seen to stop dead in the sky. The chutes collapse and disappear. There is a moment when everything seems to pause. Then the spark at the missile's base becomes as bright as the sun. The missile climbs out of the puff of smoke and away! From the distant chase plane it is silhouetted against the cloud below and beyond it, a tiny black finger rushing upward on a spark of light and a huge lumpy gray column.

In the control room there is jubilation. The men cheer at the image on the television screen. The missile is alive! It climbs on its pedestal of flame and smoke, straight up, all alone in glory between the white world and the pale blue sky.

Four seconds. Five seconds. Eight seconds. Faster, higher. Flame and power. Ten seconds.

The light flickers. The missile falters. It staggers in the air. Then it tumbles, the last sparkling thrust of its limited fuel pushing it end over end. It tumbles and goes out. The great pillar of smoke still stands there in the sky as the missile falls toward the sea and is destroyed.

Noise in the control room. Cheers. Applause. Congratulations. Big grin on Hepfer's face, a grin that looks so broad he might have a whole harmonica in there, crossways. Hot damn! We did it! Airmobile will work!

In the Pentagon any celebrations were clandestine. Toomay: "You weren't supposed to drink there. So when we got something done, Walsh would say, 'I think we deserve a party.' We'd send some guy out with a raincoat to bring in champagne and ice. You know, when you take the tops off the champagne, they shoot up. They'd make holes in the ceiling. Walsh used to say, 'If you think this is the only place that has parties, go take a look in the head shed down the hall, and look at all the holes in the ceiling.' "

Celebrations: Hot damn! We did it! Airmobile will work!

In his chair beside his pool, John Hepfer was laughing gently.

"It evidently worked," he said. "They really liked it in Washington."

II

In Washington. Michael had photocopies of the hearings he had been reading in the Library of Congress. Seymour Zeiberg, summing up the years of John Hepfer's greatest effort.

MR. LOEFFLER. Obviously being inexperienced in the area of MX basing modes, what are the options available?

DR. ZEIBERG. Let me describe what I would call the serious options which have been examined in great depth.... There was a road mobile system. That is where a missile about half the size of the MX is mounted on a special truck.... This missile and the truck, in one version, would roam interstate highways and thereby be lost, so to speak, so it could not be attacked.

MR. LOEFFLER. Put it on Amtrak.

DR. ZEIBERG. Put it on Amtrak and guarantee it will be lost.

MR. HEFNER. Or the postal system.

DR. ZEIBERG. Being roadmobile in peacetime presents horrendous safety and security problems. This, in our judgment, would never be acceptable to the Congress and to the public.

An alternate was to keep it garrisoned on military bases in peacetime and move it out in crisis times.... We estimated about eight hours of warning would be necessary to deploy the forces.... That wound up being a non-starter because of our inability to count on this degree of advance warning.

There was an air mobile concept where we dropped missiles out of an airplane. Its principal shortcoming was that it depends on warning much like the bomber force does.

MR. LOEFFLER. Are you talking about dropping missiles out of an airplane?

DR. ZEIBERG. You drop it out of the airplane and it then launches itself....

We actually did an experiment of that sort in 1974. ... That was judged to be undesireable for military purposes. Also, the air mobile concept is very expensive, twice as expensive as anything else, simply because the cost of airplanes, training, and support equipment is very high.

... [There] were a number of other concepts of land basing. There were railroad systems.... There were concepts of off-road vehicles — getting lost in the woods, so to speak.... That had awful public interface problems.... There were various other air basing concepts. One called the Seasitter, was a seaplane that would sit off the coast and take off on warning, or periodically move from one location to the other....

There were variants of commercial wide-body jets that were examined....

There were alternate sea-basing schemes such as the SUM concept, which has been popularized lately.... That is a small submarine with two or four missiles strapped on the outside.... I find it to be a step backwards, even though it has received some popular support. It has not received the support of anybody who knows the subject thoroughly....

Ideas were proposed not only to base ballistic missiles in the ocean within a few hundred to a thousand miles of a coastline, but also to base them on inland waterways such as the Mississippi River, the Great Lakes, and even one network of special canals.

[Did Zeiberg sit back and fold his hands over his belly, and smile? In the transcript it seemed as if he had.]

I could probably go on all afternoon, Mr. Chairman.

Michael sat on the patio watching John Hepfer. He had seen pictures of Hepfer smiling, standing beside mockups of successful pieces of equipment. A small man, dominated by that joyful face. Today the grin was just as big, but now there was something wistful to it.

In the distance there was the sound of a distant barking poodle. The water lapped on the edge of the pool. Hepfer looked out across the ruffled water of the inlet. He did not own a boat. If he did, he had told Michael, it would be a sailboat. He would be like Fred Payne, and sail away. He didn't seem enthusiastic about it. Those modes, those variants, those options, over which Zeiberg had so briskly and knowingly passed, had been his life. Michael asked:

"That flight was not a serious effort at Air Mobile, was it?"

"Yeah. It was not a serious effort. Although we had proposed a similar technique."

Hepfer had once succeeded John Toomay as the commander of the Air Force's Rome Air Development Center; because of the obvious comparison the men had picked up nicknames: "Big John" and "Little John." Little John, his voice nostalgic, his hands never at rest, ran lightly over the culminating years of his career.

"I stayed at Norton from January of nineteen seventy-four until I retired in November of 'eighty. Which was the longest tour I think anybody had had in that job. In nineteen seventy-four we were just about through with production of Minuteman III. They had some other program, like the silo upgrade program, that was pretty big, and the guidance accuracy improvement program, and some RV work. So there was a couple of hundred million dollars worth of work there at the time. And MX was this squirrely idea. They had, I think, about ten million dollars of study money."

Hepfer rubbed his elbow and looked out across the water.

Norton Air Force Base, 1975. Large room. Immaculate briefer. Charts. Pointers. Viewgraphs. Gentlemen, the Soviets have tested and deployed multiple independently targeted reentry vehicles on current-generation ICBMs. MIRV development and deployment is expected to continue throughout the next decade. Soviet deployment of land based missiles in hardened silos now surpasses United States totals by about one third. Gentlemen, developments and improvements in Soviet accuracy could put our Minuteman forces in danger within ten years.

Norton Air Force Base. 1975. Money beginning to move. Thiokol testing motors, bursting cases. A 92-inch-diameter missile would sure fit nicely in a Minuteman silo. It would carry ten MK 12A warheads. Draper Labs' new guidance system would do a nice job. AIRS. MX no longer squirrely, but sometimes the Pentagon seems that way. Another call from Washington: Need something right away! Got to have another breakthrough! That airdrop was great! Got to do something this year! Well, OK; how about a whole bunch of little warheads on one Minuteman? Sounds great! Call it Pave Pepper! Do it now!

Hepfer smiled and put his fingertips together. "Schlesinger again wanted it for some reason," he said. "So we shot seven RVs out there. They were so small they probably weren't much bigger than the warheads on the Poseidons. But the accuracy was amazing." He laughed. "It was just luck, you know, just one of those things that worked out. I don't think the Soviets really fully appreciated it. Their monitoring ship had been out on the range, and for some reason it had had to leave. It wasn't there when the RVs came down." Hepfer smiled, a big, friendly grin just slightly touched with sadness.

Hepfer got up and went to the kitchen to get soft drinks for himself and Michael. Michael sat listening to the water, the gentle clank of the wind bell, and the rustle of palm fronds in the breeze.

The Pentagon, March 1976. The Defense Systems Aquisition Review Council. DSARC, pronounced D-Sarc. Room 1E801. A big, noisy room full of generals and civilian brass. The D.D.R.&E., in this case the Honorable Malcolm Currie, presides. This is D-Sarc One for the MX missile. The Pentagon's first formal look. Assistant Secretary of Defense William Clements sits in. Hepfer, Gingland brief. Here's where we stand. Hard Rock Silo's dead. Rock transmits too much energy. MX missile itself looks good. Increased throw weight, improved accuracy. Options: off-road mobile, road mobile, trench, multiple shelters. Recommendation: Buy, build missiles. Put them in Minuteman silos. Continue studying multiple shelters, and hardened trench. Hardened trench looks interesting. Hepfer: "Clements was kind of a hard-nosed man." Tell your people what you want, and turn 'em loose. If they can't do it, fire 'em. Clements looks at trench sketches, and shakes his head. "Well, I know we need this system," he says. "And maybe the trench is right. But these big diggin' machines sure don't appeal to me very much."

Hepfer returned with glasses and a bottle of Coca-Cola. He sat down.

Michael asked:

"Did General Jones particularly like the trench idea?"

Hepfer put his hands behind his head. He smiled.

"General Jones —" He paused and chuckled. "General

Jones — changed a couple of times. Shortly after D-Sarc One, when Tom Reed became the secretary of the Air Force, one day Reed unexpectedly announced he was going to visit and spend the day with us. Which he did. He went back to Washington enthused, and wanted to get this thing moving and wanted to put more money into it, and the next thing I had was a call that General Jones wanted to come out and spend the day with us. We briefed him on the trench and everything."

Hepfer put his fingers together.

"He said, 'That's going to be an awful hard thing to sell.' But he became a supporter." He grinned.

Norton Air Force Base. Spring 1976. Oh, Oh. Pentagon's on the line again. Sure would be nice to have another one of those excellent demonstrations! Sure would be handy! Well, how about a new guidance system; we're about ready to test Draper's AIRS system. Great! Shoot it off! OK. Get on the phone to Draper Labs. Hello, Duff? Let's get this thing rolling.

July 15, 1976. Vandenberg Air Force Base. A flawless shot. The old king, Charles Stark Draper himself: "It looks like we've been lucky again." MX is ready to go. With the AIRS guidance system installed, MX should be able to put ten MK 12A Reentry Vehicles on target with a Circular Error Probable CEP of as little as .05 nautical miles! With 335 kilotons of power each warhead would have a 99 percent chance of destroying a 1000 psi Soviet silo!

Summer 1976. Hepfer's looking for the best men.

Hepfer has been talking to Jasper Welch. Welch says — you know Bill Crabtree? Yup. A little man. Dry. Serious. Intent. VERY intent. Hepfer was on his promotion board when he made lieutenant colonel below the zone. Well, says Welch, you ought to get him down here. He's just coming off of one hell of a job working on the A-10 Aircraft. But everybody's after Crabtree. His reputation soars ahead of him. Hepfer: "Crabtree was probably as much responsible for getting A-10 sold through Congress as anybody." Crabtree believes the way to get a program through is to have all the facts and the data, and use that knowledge as power. Hepfer likes that approach: "He did that on the A-10 program. That had a lot of opposition and he was able to provide a strong enough story that they couldn't refuse." So Hepfer goes after Crabtree. Got a good job for you at Norton, Bill. But Hepfer's bucking senior generals all over America who also want Bill Crabtree. Hepfer wins. "The Four Stars were fightin' over him so the personnel system's easy solution was to give him to me because I was junior and obviously the decision wouldn't offend them, or would offend them all equally."

October 1976. Crabtree comes to Norton. He and Hepfer assemble team. Contractors, officers. Thiokol, Hercules, Aerojet, Rockwell International, Northrop, Rocketdyne, Martin Marietta, TRW. Hepfer: "Crabtree is very astute in recognizing human frailties. Some of the people had experience but didn't have the energy or the capacity for the job. He'd just shunt 'em aside. Crabtree found a team of guys who were like he was, who would work until eleven o'clock at night making computer runs, assembling data."

MX is turning out to be the biggest single project in the Air Force. The only one for which there is truly no prime contractor, by design. Even the big companies are just associates. Hepfer runs the show. Hepfer: "When you have a prime, you never see underruns. But when the Air Force runs it, the problems are identified earlier." Hepfer, quiet, unemotional, friendly, is as hard-nosed as that Texan Clements. "That way, if anybody wasn't performin', we'd be able to eliminate them."

Norton Air Force Base. Summer 1977. Heat, smog, heavy red sunshine. People arriving by land have to be told that there are mountains out there. Heat, smog, sunshine, excitement. John Hepfer's profession by training is guidance. He was a navigator in the Second World War, and his first technical job was as the project officer on the nation's first ballistic missile inertial guidance system, designed for the Navaho missile. One of the experimental systems he worked on accurately guided an aircraft across the United States back in 1954, long before the Navaho program was canceled by budget cutbacks in 1957. John Toomay has called Hepfer, fondly, a godfather of the guidance mafia. Today this quiet, somewhat odd-looking little man, with his big ears and a grin that makes his cheeks stick out with joy, has just been promoted to major general, and is soon to direct the manufacture of the greatest guided missile on earth, the one that most elegantly combines power and precision: the new solution, the new answer to an old problem, the next invention! Hepfer's world is ready for his life's great achievement.

The Air Force is ready: Let's not wait to figure out the basing mode. Let's build it! Put it in Minuteman silos! The Department of Defense is ready: Build it! Hepfer's ready. He has been getting ready for twenty years.

But Congress is not:

Hold it. Didn't you say you wanted a safe system, not just a hot new weapon? Wait a minute.

Hepfer sipped at his soft drink. Wind rattled the palm leaves. A single mullet jumped with a splash, in flight, perhaps, from some invisible or imagined predator.

"In June when the appropriations bill came out there was a rider in there that said no money should be spent on putting them in silos," Hepfer said.

"That was the McIntyre amendment?"

"I guess that was it," Hepfer said. He put his fingertips together. "McIntyre was always against any kind of thing that would imply a first-strike capability, you know — Accuracy. Bigger missiles. More MIRVs, that type of thing. He didn't like that." Hepfer paused and folded his arms. He said mildly, "I think there's a good case for both of them, actually, but there's no question that vulnerability was an issue. His assistant up there kept harping on us that the primary objective was to attack the vulnerability problem. To get a basing system."

Norton Air Force Base, 1976. Well, OK, then: the trench. The trench! Put your missiles down inside four thousand miles of hardened concrete trench with walls a foot thick. Run them remotely around at random. If the moment ever comes to use them, jam them up through the roof of the tunnel and send them on their way. BLAM! Build the missile, put it in, and it vanishes from human eye. Tom Reed's all for it. David C. Jones is all for it! Get with it! Build the missile! Test the trench! If that doesn't work, stick it in horizontal shelters. Start deploying it in 1983! We're ready! The nation's ready!

But the Democrats have this man running for President who seems to want to get rid of nuclear weapons.

Hepfer rubbed one hand with the other.

"I read everything in his campaign," he said. "He was very clear that he was going to do away with nuclear weapons. He thought he was going to be able to negotiate this big peace with the Soviets." Hepfer shook his head. "Tom Reed was actively working to accelerate the MX operating date. Then Ford was defeated and Carter came in and I knew we were doomed to start all over again, trying to explain what we were doing."

The wind rattled the palm leaves and rang the clay bell. No sign of remembered dismay escaped from John Hepfer. Nor did he smile. He looked out across the water. He said, with mild interest:

"Have you ever seen these pelicans dive?"

14

I

SUMMER was approaching. News clippings piled up on Michael's desk like falling leaves. "Meese says Reagan may relocate MX." "MX Plan Blowing Holes in Conservative Ranks." "Two influential Republican Senators, Paul Laxalt of Nevada and Jake Garn of Utah, announced today that they would oppose the deployment of MX mobile missiles in their home states." "Senator Tower Assails Proposals of Garn and Laxalt on MX Basing." "Panel Considers Compromises on MX Basing Plan." "MX worries impel Boeing alternative." "Interior Greases Gears In Commitment to MX."

The Baltimore *Sun* reported that Utah faced "a decade of disaster" as the MX project and energy development flooded the state with new population. The Washington *Star* reported that a committee formed by Defense Secretary Caspar Weinberger to recommend changes in the MX system was thinking about reviving an anti-ballistic missile defense system instead of hiding missiles. United Press International reported that the Air Force was suggesting a smaller version of its multiple protective shelter system as a compromise plan. Columnist Russell Baker reported receiving a letter from the White House asking him to explain his MX Pentagon system. He also reported that a prototype of his mobile Pentagon had been hijacked in Indiana shortly after a collision with the town of Terre Haute. Stories in the Washington *Post* and the New York *Times* reported that Secretary Weinberger had called for a United States military capacity to carry on two simultaneous wars: ". . . we must be able to contend with a conflict in one area without opening up critical vulnerabilities elsewhere. . . . We will not restrict ourselves to meeting aggression on its own immediate front. . . . We must be prepared to launch counteroffensives in other regions and try to exploit the aggressor's weaknesses wherever they exist."

Michael took a long afternoon walk in the hills near his home. The weather was unsettled. There were puddles in the road. Tall clouds blew across the sky. From one of them hung a gray wall of rain. As Michael glanced west the clouds moved and the sun, hidden from Michael beyond the shower, emerged to light the rain. The rain turned a brilliant white. Michael stopped and watched. The shower detached itself from its cloud and fell, still bright, to the ground. It splashed brilliance all over the fields below. Michael stood on the road while the cloud moved toward him and the sunlight vanished. A gray shower again developed, but Michael still watched the farmland. It had seemed as if a loose-knit piece of light itself, a swarm of light, had fallen through the clouds to caress the land.

Michael stood among the puddles on the road staring out across the sky and the land until the rain reached him. It swept over him in a fierce downpour. Still he stood, looking out, until the curtain of rain slowly washed the entire scene away in grayness. The rain was icy cold.

II

"Tell me about these people."

Michael glanced carefully at the woman across the table. She and her husband and Michael were having lunch. The couple was active in the anti–nuclear weapons movement. Her husband had told her what Michael was doing. The Boys behind the Bombs. Her face was bright with respect.

"Well," he said. "They're . . . They're very different from one another. There's John Toomay. General Toomay. He's a remarkable man. Forthright. Very forthright. You ask him something and he sort of takes it, digests it, then bursts out with whatever he thinks of it."

She was interested. She had earlier been talking with anger about what she called the Pentagon Mentality. Michael continued.

"The first time I met him, there I sat in his home, total stranger. I asked him about the MX history. He thought for a minute, then he said: 'For several years the Air Force was going down the road to a tunnel. When any thinking person could see that the tunnel was stupid.' This is an Air Force general. Toomay is just constitutionally unable to be coy."

His friend's wife was looking at him with a vague perplexity. He rushed on.

"Seymour Zeiberg. He's not that way at all. Always a little withdrawn, a little superior, but not unlikably so. A kind of raucous sense of humor. You feel him being — pompous without being cold. He comes on as both vulnerable and — arrogant. Constitutionally unable to be — to appear ignorant. Then there's Albert Latter. Pacing around. Staring out his window. Constitutionally . . . restless. And Marvin Atkins. Hard man. Not easy to know at all. We sit there in his office, and we laugh and laugh, and we're both as hard as iron."

She looked as if she expected more.

Michael went on.

"One of the guys I haven't been able to talk to is the MX project engineer, a guy named Crabtree. I've spent some time with his former boss, General John Hepfer." Michael paused, hearing the ripple of water and the rattle of palms. "John Hepfer," he said slowly. "What a wistful man."

His friend's wife looked at Michael coolly.

"You seem to actually like them."

He looked at her. He said nothing.

III

From his home, Michael called Neil Buttimer. Buttimer was out.

His assistant seemed to be the sort of person to whom everything is catastrophic. Colonel Crabtree was away, she said hopelessly. She didn't know when Crabtree would be back. She didn't know when Buttimer would be back. In a voice of apparent desperation she gave Michael the number of a colonel named Larry Molnar.

"Why don't you call him up yourself," she said, much moved. "He'll probably talk to you."

A few days later Michael drove slowly into Norton Air Force Base. The guard at the gate was a woman in fatigues. She paid no attention to him. He passed a large sign: "Strength Through Vigilance." He drove between rows of impossibly tall and slender palms. They looked like creatures from a lighter planet. The lawn looked too green. He drove over to the Ballistic Missile office. The Minuteman missile outside looked as if it was turning gray. A soiled obelisk. He looked up at the Minuteman symbol on the wall: the Revolutionary War soldier with his rifle, seeming to lean in his angry zest just a little too far forward.

Michael found Larry Molnar, with the help of an escort, in an office off a labyrinth of halls. Molnar was large, gray-haired, broadly cheerful. On his desk was a mug with "MX" on the side. On one wall was a montage of color photographs: huge machines with rotating toothed buckets, ditches in pale brown earth, a hydraulic arm poking up out of the ground bracing a yellow cylinder. Molnar was the man, Michael had been told, who remembered the testing of the trench.

Michael listened to Molnar. Molnar seemed oddly familiar. There was something recognizable about him. Molnar leaned far back in his chair. His hair was swept straight back from his large face. It made him look as if he was moving ahead with all deliberate speed through a viscous liquid. He spoke the same way, pushing along in a careful, steady cadence, marked by a rhythm of emphasis.

"That site was located in *Stovall*, Arizona, which is, to my recollection, not an *inhabited* city. It is an *abandoned* airfield. It may be *difficult* to *discern* precisely what was *accomplished* at that site."

Molnar's was a story of heat. "I went out there many *times*," he said. "And there was no *doubt* in my mind that the *heat* did not really affect the *personnel* in any *adverse* fashion. Those *people* who worked on this *project* are all *natives* to that part of the country. The *operation* started out at *six o'clock* in the morning and by about *two o'clock* in the *afternoon* people were actually *leaving* the *site* before the hotter temperature of the *day*. But they were certainly *experiencing* some hot *temperature*."

At Molnar's suggestion Michael turned to the photographs on the wall. He stared at the montage.

Southern Arizona, east of Yuma. A cluster of trailers, trucks, and outlandish digging equipment on that heatblasted desert. Hot men. Acclimated men. Men working in sweat-stained hats and sweat-stained T-shirts, and their faces burned by the reflection of the sun from the pale desert floor. Men digging a twenty-foot-deep trench in the dirt almost three miles long. Men filling the trench with a concrete tube.

The tube is seventeen feet in diameter. It emerges, like some enormous piece of gray feces, from behind a complicated machine that looks like a skyscraper's scaffolding that has become all crumpled and compacted after having fallen into the street, and has then been partially shrouded in sheet metal. A stream of concrete trucks rushes past this machine, each truck depositing its contents in an aperture, then rushing back down the slope for more. Men clamber over the machine's misshapen surface. Far ahead of it — in time or in distance — two multibucketed trenching machines gnaw at the earth, spilling it in cascades to the sides. One cuts a flat-bottomed notch ten feet deep and the next goes down ten feet more, leaving the bottom concave. Men climb all over the trenchers, changing carbide teeth. Ahead of them roams the finest land minesweeper in the world, picking out buried fragments of 20-millimeter shells.

Around it all, tanker trucks filled with water that has been pumped eleven miles drive aimlessly up and down the dirt, sprinkling. The water almost hisses as it touches the earth and evaporates.

For miles on all sides of this tiny patch of activity, the desert spreads: empty, dry, silent, infernally hot.

"It was *interesting* how we had to produce the *con-*

crete," Colonel Molnar said. "We had a concrete *batch plant*, near the runways, and an *essential* part of that batch plant was an *ice machine*. In the *manufacturing* of the concrete, which was very much *controlled*, they devised a *formula* so they could *mix* the proper amount of *cement* with *fly ash* and with the gravel and then *prior* to *moving* it from there, they also used *ice*, so that while it was being *transported* to the *site*, the *texture* of the *concrete* remained the *same*."

Michael turned again to the photographs. They dominated the wall, beautifully mounted, like mementos of true achievement. He stared at them.

Summer 1978. Sun, heat, shimmering clear air. The only thing stranger than this desert scene is the eventual purpose of the concrete tube being built there. When it is done, the tube will snake up the side of the broad valley toward a black hump of mountains called the Mohawk range. The tube will lie in at least two sections under five feet or more of earth, and will have several entrances, which will consist of graded earth ramps leading to enormous steel doors. It will be possible to open one of the doors and drive a missile inside on a machine called a Transporter-Erector-Launcher, a TEL. In the top of the tube will be two holes, filled with enormous hatches. At appropriate moments these hatches will be removed so photographs can be taken of the underground TEL — from space. At any moment the TEL will be able to stop inside the tunnel and, upon command, push the missile in its canister up through several feet of concrete and dirt so it leans on its hydraulic arm at about a fifty-degree angle pointing northwest — BLAM!

The TEL is called the Strongback. It is a logical name for this odd, powerful piece of equipment, that can raise its muzzle against such enormous weight and structure, like a mighty cannon, and fire its projectile one-third of the way around the world: Strongback.

Michael watched Molnar. Why was he familiar? Molnar went on.

"We did write an *impact* statement and hold a *public* hearing. There's a *large* high school, which is in *close* proximity. It's at a *place* called *Rolle*. We had a public *hearing*. I wouldn't say we had *large* attendance. They *came* and they listened, they asked very few *questions,* and then we *filed* it. We *didn't* have any *adverse* people to that project."

"If sometime — if I were to go —" Michael asked. "If I went down there to Arizona — could I see the trench?"

"There is a *door* on it," Molnar said. "And the door is *locked*. I'm not sure that *many* people have the *key* to that." He laughed. "We really don't want somebody to *go* in there and — you *know*, it's a very good place for *animals* to hide. It could get to be *dangerous* in there."

Michael asked:

"You must have had people coming out there during that decision-making process."

"We had *some*, yes we did. The construction *project* wasn't as *visible* to people in *Washington* or people outside our ballistic missile *office* as I think we would have *liked* to have had it been."

Michael replied, unconsciously volleying the familiar emphasis back:

"I *see*. It was something that was going on *out there*."

Molnar returned, neatly: "It was something that was *going on* out there. They would say, 'Well *is* it going on?' Southern *Arizona*. Then they would say, *'Well,* ah, when are the major *activities* going to *occur?'* July or August. They'd go, 'Um, that's got to be the *hottest* time of the *year,'* and — it was not that *interesting*."

"Now, which year were you doing all this?"

"It was 'seventy-eight," he said, uncertainly.

"Couldn't be 'seventy-eight," Michael said. "Why in 'seventy-seven —" He caught himself, and proceeded cautiously. "It seems more like it would have been in — in early nineteen seventy-seven."

"Well, 'seventy-seven sounds *reasonable*," Molnar said. "But the only thing is, *jeeze*, in 'seventy-seven I had only been here for six *months*. Could have been, though."

Molnar called Neil Buttimer on the telephone. Buttimer was in. Buttimer brought a copy of the trench's Environmental Impact Statement with him.

"Oh, well," Michael said, with Buttimer standing in the door. "It — ah — doesn't *seem* all that important."

Then Buttimer left.

"Must have been 'seventy-eight," Molnar said. "The public *hearing* was held in *September,* nineteen seventy-seven. So we started work in 'seventy-eight."

"Amazing," Michael said. "So before you even started —" But again he didn't finish. He was looking at his notes. By the end of nineteen-seventy-seven, he had written in notes made in the Pentagon, the Department of Defense had already made up its mind that the trench was dead.

Michael said nothing more about it. Molnar, without prodding, had suddenly gone into a philosophical speech.

"The *bomb* apparently has changed *everybody's* perspective on —" He hesitated. His voice had deepened, and gone hard. "— *That's* what I get out of reading *books*. I think there's *way* too *much* analysis done on why we — the *psychological* aspects of *war,* and a few things like *that,* instead of *recognizing* the cold *facts.* We *invented* the bomb. We've known how to *control* it for a *long* time. Now the Catholic *Church* comes out and says, hey, don't put the next *generation* system in *Utah,* that's *bad.* What the Catholic *Church* should really preach is the fifth *commandment,* you know, to take care of the eleven hundred *guys* killed here in Los Angeles with *guns* every year. Plus the fifty-thousand *people* that get *mutilated* on the highways. You know they *just* found a kid's *arm* the other day. On the *highway,* see."

Michael just watched him. The familiarity was the way Molnar spoke. Molnar sounded just like an Army drill instructor. His was a caricature of military diction: Crisp, authoritative, emphatic. Those sergeants had talked the same way: "Insert the *round* into the *chamber.* Just like you *insert* your *cock* into her *cunt.* When you see that *gook* running you want to knock him *down.*"

Michael looked at Molnar, his eyes large as a cat's. He sat rigid. Molnar leaned comfortably back in his chair.

"Another *thing* that you're not going to get *away* from. You know there's this big *prime target syndrome* that everybody has, like as if they're *only* going to become *prime targets* now, see, with the MX. The *Russians,* if they're going to win the *war,* are going to have to *destroy* the *warmaking* capability of this *country.* A lot of *people* think the warmaking *capability* of this country is its tanks and its MXs and its *Minutemen,* and all the rest, and they're *wrong.* And that's being *proven* now by Reagan going *ahead* with the *neutron bomb.* What's the *neutron bomb* kill? Kills *people.* See, the *warmaking* capability of this *country* is in the *Army,* and the *Navy. The Air Force. It's* in city *administrations,* county administrations. *It's* in the *people.*"

IV

Michael left the Air Force base on Tippecanoe Ave., headed south. He passed the building that had once housed the Aerospace Corporation. Somewhere down there had been the famous restaurant, Pitrozello's Del Fino de Oro. The place to go now, Marvin Atkins had told him with a laugh, was Bobby McGee's Conglomeration. When Michael had eaten there he had found a place festooned with stained glass and old etchings, where tarnished musical instruments hung from the ceilings and the waiters and waitresses dressed as Robin Hood, Li'l Abner, Apollo, or Belle Starr. His cocktail waitress had been tall, slender and clad in a tailored sheet. She looked severe; unlikely to go astray. Her name, she had told him seriously, was Athena.

He drove on.

The countryside around the base had the look of debris: convenience stores on empty blocks, isolated bars, empty desert lots glittering with glass, bedraggled brush, roads that widened meaninglessly into avenues then shrank again, sidewalks that appeared and disappeared, afternoons cloaked in an impenetrably ruddy smog into which the sun sank like a bruised plum. It made it seem as if Norton's surroundings had once been neat little subdivisions, treed, clear, cool, and that suddenly one day this infernal hot wind had arizen and, bearing red dust, had blown most of it away.

Michael spent the night in a motel on the edge of San Bernardino, near the intersection of two highways. He lay in bed in a room full of moving shadows, staring at the ceiling. The highway was busy. The walls of the motel were thin. The little room glowed through its curtains and shook when trucks passed. It was as if the room itself was traveling. It was as if the room was a wood and cardboard case on a flatbed, traveling out of the familiar and into the permanently strange. Michael heard Larry Molnar's voice again and again. It's in the *people.* It's in the *people.* It's in the *people.* The room traveled. He lay awake.

15

I

ALBERT LATTER crossed the room, and stared out into a dazzling Southern California day. For a moment he looked like an ascetic commanding general pacing the floor, waiting beside a silent radio, trying to peer over the terrible horizon. Michael watched him. Here again was that same unquenchable restlessness, that had sown a seed for MIRVs, that was never satisfied with the known fact or the accomplished work. That white hair swept back, that pronounced, bent nose, that urgent stride, those searching eyes.

"Teller had asked me to talk," Latter said. "He felt that what the Air Force was doing might not be the best thing to do. I presented my criticism of the so-called trench system. I said, 'Don't do it that way, do it this way.' "

He was talking about a meeting of the Air Force Science Advisory Board, a panel charged with evaluating and critiquing the Air Force's technological plans. The board was an important organization; among its members were Dr. Edward Teller, known as the father of the hydrogen bomb. The meeting had been held in the late fall of 1977, about the time Larry Molnar and his crew of contractors were planning their excursion in the desert.

Latter stared out at the gleaming Hollywood hills. His small eyes squinted, at either the glare of the light or a sharp memory. He said:

"There was a certain amount of haranguing."

Haranguing! Michael too squinted, and followed Latter's gaze. Norton Air Force Base, the end of 1977.

The Carter administration is now in command. Carter has chosen Harold Brown as his secretary of defense. Brown, a shy, brilliant man who prefers documents to briefings, has chosen William Perry, a shy, polite mathematician who has run a military electronics laboratory for ten years, as his undersecretary for Research and Development, a job with a new title, still called by the old one, D.D.R.&E. Bill Perry, who took the job reluctantly, has chosen Seymour Zeiberg to be his deputy for Strategic and Space Systems. Zeiberg was not reluctant. He has brought his gong and his vigorous crewcut to the third floor of the Pentagon. Zeiberg has chosen Marvin Atkins to be his deputy, in the position of Director of Offensive and Space Systems. Atkins has brought his laugh and his questioning eyes and his little brown cigarettes. John Hepfer and Crabtree are at Norton, two small men of drive, Hepfer with his grin, Crabtree with his relentless intensity. Big John Toomay is at Systems Command. Albert Latter is driving out to Norton to take on the trench.

It is a pivotal meeting. The trench is about to take off, in its lumbering way, and echoes of this meeting, in support or criticism, will be heard through the Pentagon. David C. Jones, head of the Air Force, likes the trench, and who is to stand in his way? Albert Latter? This playful person on the outside, whom people can not ignore because although he may seem iconoclastic he is damn smart. He swings a heavy length of chain. So heavy, and long, in fact, that Marvin Atkins, a friend and colleague of Latter's from way back, has flown out to Los Angeles just before the meeting to try to tone down his criticism.

Atkins: "I went out there a day early because I had spent at least three hours on the phone the previous two or three days with Albert, arguing with him. And I finally said: 'Albert, I can't take this anymore; I'm just going to come out and see you.' "

Atkins, hard, lean, suspicious. Latter, hard, lean, intense. Decoys. Costs. The trench. Atkins: "Some of his ideas were unrealistic. He wanted to build perfect decoys." But Albert, that's just impossible! "He got himself in a lot of trouble by having an exceedingly flaky cost estimate. He didn't like the trench."

Atkins: "I spent the day trying to modify his extremism so he wouldn't get totally shot down in flames. I would say I was only partially successful."

Norton. A room full of scientists, brass. Hepfer, Crabtree. TRW people. Albert Latter pacing before the scientists, his colleagues and critics, drawing on a blackboard, his eyes darting among them, staring, searching. Searching for — what? "Do it like this": A line on the board, marked 7200 feet at the right end, 3600 feet in the middle, zero at the right. Even if he doesn't know where it is, all he has to do is put a bomb here (halfway between the 0 and the 3600) and here (halfway between the 3600 and the 7200) and he has it. So you don't need a trench. Just put shelters at those points and connect with a road. The shelters are as good if he doesn't know where it is. You're paying for 3600 feet of nuclear hardening. It's a waste. With horizontal shelters — garage mobility — you get two modes of protection — Dash on warning; and deception. Good to have two modes, because you don't ever know how the political people will respond.

Albert Latter turned from the window. He sketched on the board. He came back to the middle of the room and shoved his hands into his pockets. He said:

"My recommendation was: One, get rid of the trench. Two, if you are willing to live with deception only, go to vertical silos, because they're safer." He sat down. "Three, use small missiles, Minuteman size, not MX size. And four, if you decide to be dependent on deception, you must have decoys."

Norton. Decoys, decoys. But ALBERT, it's just going to be the greatest game in the Air Force for all those NCOs and privates to guess which is the real missile and which is the dummy, and there is no way in the world that you can stop them from finding out! It's going to be a big game and they're going to make bets against each other, and they're all going to know which are the real missiles.

"Latter was irrepressible," another visitor to that meeting had told Michael. "They dealt with him very harshly. I have seldom seen such poorly veiled sarcasm. Why don't you sit down, now get someone else. He just hung in there; he said the only way to go is multiple silos."

Ivan Getting, the crusty head of the Aerospace Corporation, is chairman: Albert, you've gone on enough. Glenn Kent, a three-star general, former head of Air

Force Studies and Analysis: Albert's proposal is good. You're just rejecting it because of the way he's presenting it. Kent, later: "Albert has a tremendously fertile mind, a brilliant mind. But he is the last person I would send out to sell anything. Albert has a way of introducing a subject; before he'll tell you what his proposition is, you have to admit you're stupid for having thought of anything else. Albert was so busy telling people how foolish they were to have thought the trench was a good idea, that he obscured the utility of his proposal."

Restlessness, anger in the group. Cold eyes, cold faces. Teller. Art Beihl. Getting. John Hepfer, no grin. Bill Crabtree. Glenn Kent, a wiry, tenacious man, with eyes hidden in slits of weathered flesh. Glenn Kent, a man with a face as mobile as liquid. Kent likes to flex his understanding of the English language. Get that participle back where it belongs, soldier! Latter: "If Glenn Kent is there there is sure to be some question about the punctiliousness of language." Atkins watching with sardonic eyes. Invisible in the background: the uncertain shape of the biggest Air Force project. Tunnels, concrete, diggin' machines, a 190,000 pound missile, the deadliest weapon in the world, the new invention. In the room, an array of icy faces, with Latter a windblown flame among them.

Michael, talking to Glenn Kent: "So here you are in this room talking to each other?"

Kent: "No, no, talking BY each other."

"Is all this feeling muted, or does it become heated?"

Kent: "Yes, yes. It becomes heated."

At last agreement is reached. Albert, Sit down. You're done. Be quiet. Sit there next to Glenn. You've talked for three hours, and we're still not sure what your proposal is. Glenn, you seem to feel Albert's ideas have merit. Would you please represent them to us? Albert, you may pass notes to General Kent if he gets anything wrong. Kent: "Obviously they weren't totally able to enforce that discipline."

Scientific advisory boards don't make decisions. They validate or cast doubt. This one casts a long uncertainty, although that day, when it ends, it seems nothing has changed. It breaks up in cool amiability. Later, John Hepfer takes Albert Latter to lunch. Latter tells Hepfer that maybe sometimes he exaggerates a little; it's just a habit left over from the days he wrote radio scripts after the war. "The Falcon." "Inspector Burke of Scotland Yard." "Exploring the Unknown." Got to catch the audience. Hepfer offers a grin of respect: "Latter was a thorn in your side when you were really trying to get something done. But he made us look at a lot more detail, which was not a bad thing."

Kent: "Albert was right. He did a very useful thing that day. That was the beginning of the turning away from the trench. People had already had misgivings, but all of a sudden he was the catalyst. Even though it was a rather — intense — meeting. You can imagine that it was a little hard to take."

Atkins takes a photocopy of the uncertainty back to the Pentagon. The Air Force's next move is to send its director of concepts and analysis, Office of the Deputy Chief of Staff, Programs Analysis out to California to have a talk with Albert Latter. The major general who occupies that position is a nuclear physicist: Jasper Welch. He and Latter go skiing.

Michael watched Latter pace the room. The light outside the window was too bright. It hurt Michael's eyes. "We went to Mammoth," Latter said. Michael knew the road. Three hundred miles. The last time he had traveled it had been at night, in a full moon in winter, and the mountains had shone, white on dark, like photographic negatives. Now he could see them traveling in the same half-light, riding in a luminous netherworld between weapons and war.

December 1977. Jack Welch and Latter have known each other a long time. They have similar minds. Like Latter, Welch got his Ph.D. in physics at Berkeley. He worked with Latter at the Rand Corporation in 1962. Welch tends to let his uniform get sloppy, and sometimes others aren't sure they know what he's talking about, but he gets things done. He has a lazy, self-confident face and eyes that appear drowsy but never sleep.

Toomay: "Welch's ability to organize an analysis is remarkable. A remarkable talent. Jasper Welch is a guru."

So in this winter of 1977 those two old friends, the one now a two-star general, the other head of a noted research and development organization, drive the long narrow road from Los Angeles to the skiing village of Mammoth. With another friend at the wheel, Welch in the front seat and Latter in the back, they rumble along U.S. Highway 395 below astounding mountains, through little communities, past dry farms and lava beds, talking about the MX.

Lancaster, Mojave, Red Rock Canyon, the turnoff to Edwards Air Force Base; the turnoff to the naval ordinance station at China Lake. Inyokern, Little Lake, Haiwee Reservoir. Latter's eyes squint at the black rock, and the windblown glittering water, hungry. If you draw a line, and you put zero at one end and 7200 at the other . . . If you build a road, and you put shelters on

it, at 3600-foot intervals...How hard can you build
a silo, a missile shelter? How hard does it have to be?
Can you be sure of 3000 pounds per square inch, or
must you conservatively estimate 2000? If so, how far
apart must they lie, so that when he throws a bomb
halfway in between he cannot get them both. Filled
hexes. Spike attacks. If you draw a line —

Olancha, Cartago, Bartlett, Lone Pine. Mt. Whitney,
there, highest in the nation. Shining mountains. Hunger.
A long, straight road. Ah, Albert, you misunderstood
the political solidity of the trench. In due course the
trench will wind up being horizontal shelters. We know
it. It has to work out that way. It is perfectly silly to
buy expensive roadways for parts that don't need to be
expensive. Dash, deception. You need the two modes,
Jack. You never know what the politicians will go for;
offer them both.

Lone Pine, Independence, Big Pine, Bishop. Grandeur.
The loft and pomp of the Sierras, to the west; the bleak-
ness of the White Mountains to the east. How far apart
are those shelters going to have to be? Three miles,
Jack! One mile! Three miles; there's a lot of land out
there. Silos have never been tested in a totally realistic
environment. Arctic Night; damn test ban; three miles!
One mile? If he puts five megatons a half mile from
each, are you going to guarantee strength? But, Albert,
you're going to fight for every inch of that land.

Owens River, Tom's Place, Mammoth Lakes. Fifteen
feet of snow. Tunnels to the condominiums. Trench!
You're going to have to close off all of Arizona to the
public so somebody doesn't get in there with his ear to
the ground! What if he hits just one end of that tunnel?
That shock will roar down that acoustic tube like the
charge of a shotgun, and blow your missile out the end.

That's not the way to launch it. And how are you going to get the public to pay for that tunnel at a million and a half a mile? Albert! Relax, Albert. We're going to do something else. Now look, Jack . . .

Mammoth Mountain. Albert Latter, easing down the mountainside in sweeping, graceful turns; standing in the lift line, looking up the mountain, searching.

Latter stood at the window. The light outlined his raptor's profile, his eyes slitted as he looked into the city. "Then," he said, "I met with General Lew Allen." Latter walked back and sat down on a couch underneath the picture of the racing skier, flexing a wooden pointer between his hands. "He at the time was head of the Air Force Systems Command." Latter got up, walked over and sat down in a chair on the other side of the room. "Allen was persuaded," he said. "But you can't force the Air Force to change the system in such an informal matter." (The trench is crazy! OK, let's do something else.) Latter got up and looked out the door. Then he came back to the couch and sat down.

Latter's windows blazed with light, making the view an indistinct glare. Michael watched him, his eyes wide. What was out there, so bright, so important, so precious? Michael seemed to feel the air in the room moving, surging with the passage of someone in an adjacent hallway. He watched Albert Latter. Cool hands, eyes exploration itself. The physicist. The inventor.

Jasper Welch: "Albert has played a particularly interesting role in most of these things. In about three or four occasions he has sort of entered into the decision process with a highly placed visit or letter and been very effective at disrupting the decision process. He's always had good and useful things to say. And I adore him and highly respect him. As does Harold Brown and, indeed, Lew Allen. I talk to Albert every month or so. In those days it was quite often. In fact if I remember, he sent me a draft of his letter to Brown which I commented on. So there was that kind of closeness.

"But his timing is just either exquisitely good or bad depending on your point of view. He was good in the sense that he entered into the high place just about the time those people were engaging in the process, but of course it was coming just at the end of the Air Force's preparations, which made it extremely hard on everybody else."

So now Latter squinted and his eyes searched Michael's face — for what? "My suggestion to Allen," Latter said, "was that he had a good general there in the Air Force systems command — I suggested that he form a committee, and that a good guy to head it up would be John Toomay."

Latter stood again and walked toward the window, and to Michael's dazzled eyes he vanished into the light.

II

The telephone rang.

"Hello. This is Neil Buttimer."

"Colonel Buttimer! How are you?"

"Fine. How does nine A.M. Tuesday sound?"

"Nine A.M. Tuesday?"

"To see Colonel Crabtree."

16

I

MICHAEL drove through the early morning on the Southern California freeways. The sky was clear. Buildings stretched from horizon to horizon. Michael passed familiar exits. There was the road to the Santa Susana Mountains, where the Rocketdyne company tested fourth-stage MX motors among sandstone boulders and desert scrub. The roar could sometimes be heard in the San Fernando valley. When Michael had visited the test grounds the fog had been thick, and the little test stands had loomed strangely from among enormous round rocks. He had been escorted around Rocketdyne by a pleasant public relations woman who wore a big ring on which was the model of the head of a Doberman pinscher.

There was the exit to Marina Del Rey, to the tower on Admiralty Boulevard above the water where Albert Latter worked. There was the turnoff to the headquarters of TRW. At TRW Michael had sat in a quiet office and listened while a vice president had told him: "There are two groups among the men who invented the bomb. These are men who know what the effects are. Some of them went to Hiroshima to study the effects. Both consider they have original sin. Both groups never want to see it used again. One group wants to wipe the bomb from the earth. The other group, typified by guys like Al Latter, thinks no cost is too great for strategic stability."

On the left now was the Aerospace Corporation exit, the road to the plant that had replaced the San Bernardino operation in 1972. Michael had spoken there with a vice president who had told him: "Seymour Zeiberg was like my younger brother, my successor, my protégé. In the beginning what Sy and I were doing research on was fundamentals. The fact that this thing carried a nuclear warhead probably had zero motivation at that time."

Farther along was the road to the Hawthorne Airport, where the Northrop Corporation had a headquarters and a plant for the manufacture of MX guidance packages. Michael had been shown through that plant by a vice president who strikingly resembled Robert Duffy, that short, ebullient, pipe-smoking president of Draper Labs who had lost the Athena, even more in enthusiasm than looks. In the Northrop lobby was a painting of the MX, with the legend: "The United States Air Force's next generation ICBM, called MX, will be guided to its target using a Northrop-built 'floating ball' inertial measurement unit." In the painting the missile flew against a background of fire.

There was a turnoff to the north that led to the offices of the Rockwell International Autonetics Strategic Systems Division, the people who had told Albert Latter back in 1963, yes, we can build a MIRV bus, and we're now doing it for MX. There Michael had seen a promotional slide show of Autonetics products — missile guidance, battlefield sensors — that had involved twelve slide projectors and a strobe light. The show had begun with the somber words: "John Milton told us of that first war . . ." and had included a photograph of a heap of skulls. The show was brand new, and someone had asked him afterward if it sounded too warlike. He had replied ambiguously, but on the way out of the room he had overheard an answer from an Autonetics executive who was talking to a colleague: ". . . said there was too much about war. I told him, after all, we don't make . . ." On his way out, Michael had thought he had heard the man say, "Broomsticks," but he was never sure.

And there was the San Diego Freeway, heading south to the small town on the coast where John Toomay lived.

Michael drove on and on, flowing with thousands of other people in other cars across a landscape dominated by shelters for human beings in which these pockets of MX construction hid. But the open sky dominated the world, and under the sweep of mares-tails clouds and blue the endless city looked fragile, just a dusting of life, so easily blown away.

The mountains north of Norton Air Force Base were in the clear. They stood on the horizon, grey-blue, with a sense of mass behind them, like a retaining wall. Michael drove the familiar roads between the palm trees and parked in the Post Exchange parking lot near the weather-stained Minuteman that stood in front of BMO. Crabtree's office was near the front of the building.

Michael sat at a long table in a large room. The room seemed too low, and the table seemed too long. The stack of papers at the opposite end of the table looked so heavy that the table should have listed in that direction. On top of the stack was a box that looked as if it contained a board game. On its side was printed the word "Ultimatum."

Neil Buttimer sat on the other side of the table, his smooth face oddly powerless to spread boredom, as if it had been overwhelmed. Next to Michael sat Colonel William Crabtree. Michael's eyes were cold.

Crabtree was a small man with a round, youthful face a little broad in the cheeks, with a flash of gray at the sideburns. The gray looked out of place. He was strangely supple: He was slumped in his chair, and had hooked one leg over its arm. He looked almost like a contortionist; it might not have been surprising, or much more of a twist, for him to have hooked the heel of that glossy uniform shoe around behind his head.

Crabtree did not stare Michael resolutely in the face, like Toomay; he did not search Michael's eyes and the windows and the corners of the room with equal ferocity, like Latter: his gaze wandered, ducking intimacy. It glanced off the wall, sidled over to Buttimer, nodded at Michael, touched on the stack of papers on the table, checked the shine of Crabtree's own shoes, then flicked over Michael and to the wall again.

Crabtree's voice was a monotone, and his words were delivered in short bursts, steady as the sound of a telegrapher sending Morse Code. He sounded as mechanical and emotionless as a human being can sound without speaking through a length of pipe.

"I came to the process in nineteen seventy-six — when the baseline for the ultimate basing system — was the hardened buried trench — or tunnel. At that time it was also the closing days — of the Ford administration — under the leadership of Secretary of the Air Force Tom Reed — there was also an effort — to put this missile in the current Minuteman silos. I came from — five years in the office of Studies Analysis — I was pretty well a card-carrying analyst. The first piece of work that we did here was at the direction of the Secretary of Defense — in which he asked us to look at the — possibility of putting the existing Minuteman missile in the trenches. That work was done in the fall of 1977."

Buttimer sat like a sphinx; Crabtree talked like a piece of computer hardware designed down at Rockwell Autonetics. Michael stared at Crabtree with the eyes of a cat spotlit on a prairie road. Outside the room he could hear the sound of typing, and voices; he seemed to hear the whole vast acreage of the buildings of the Ballistic Missile Office — those seven interlocked warehouses, maze within maze, room within room, photos upon photos of tall, slender objects blazing fire, BLAM! BLAM! BLAM! — murmuring with the flat, stacatto directives issued from this office, the heart of the MX.

We will look at the trench — and there goes Larry Molnar, two hundred and twenty men, and an assemblage of hardware never seen before or since on the face of the earth, off to Southern Arizona. We will look at the trench — jobs occur, careers move upward and downward, piles of gold and paper disappear from the federal treasury; the trench appears — and nobody goes and looks at it. And Colonel Bill Crabtree goes on:

"I actually came here — as project officer for the post boost vehicle — the primary missile. After a few months I became — the project officer for the entire missile — it was clear at that point — there was going to be a lot more studies — before the MX decisions were ultimately made." Michael watched him.

Norton Air Force Base, 1978. Bill Crabtree at work: Albert Latter (and others) suggest that maybe the trench would carry shock like an acoustic tube — or a shotgun barrel, and will blow the missile out the end — so Crabtree's office works on Blast Plugs. Run a giant cork up and down the trench at each end of the missile, and block off the shock. Or, build spurs — little sidings off the trench at right angles where the missile could duck, like a mouse into a hole, when the blast runs its ugly long paw down the passage.

Oops — the White House is worried: What happens if the trench is such a good hiding place that you can't tell whether there is one missile in there or two dozen? Then all your treaty agreements on Strategic Arms Limitations will be jeopardized. So Crabtree's office works on SALT Ports, hatches in the top of the trench which can be opened periodically to let Soviet satellites peer in. Eventually he works on a variation of that idea in which the entire sixteen miles of each trench — still strong enough to withstand a near miss of a one megaton bomb — can be opened to display its contents, like an artery parting its lips to reveal its race of corpuscles. That one is officially called the Openable In-Line Hybrid Trench. It is known as the Zipperditch.

But, ah — maybe the trench will not be a good enough hiding place. What if the Soviets can send a plane overhead, or a spy on foot, who, with his fancy equipment, can detect the shudder the earth will make when the weapon slides beneath it? So Crabtree's office briefly contemplates a device that can simulate that subterranean rumble — which becomes known, in some circles, as the Ground Thumper.

Day and night Bill Crabtree works with his team, sorting out suggestions from secretaries and undersecretaries, Air Force generals, Navy admirals, independent scientists, and contractors, feeding them into his computers, flying back and forth to Washington, briefing secretaries and undersecretaries, building a program, building a reputation. Everyone knows Bill Crabtree.

Hepfer: "Crabtree was this way: Bill Perry would come to visit, spend three or four days. When he left one evening he would give us a list of actions he felt we should take. At seven o'clock the following morning Crabtree would have all the engineers and project officers in and he'd have briefing charts of exactly what he wanted done."

Marvin Atkins: "Crabtree is very intense. He is a very bright guy. He really does have an emotional relationship to the work. He is widely respected by everybody who knows him."

Hepfer: "Bill Crabtree probably briefed about everybody in Washington at one time or another."

Kent: "Crabtree is an extremely good engineer. He has a grasp of many different subjects. He was totally overworked at BMO. Every new problem that came up went to Bill Crabtree."

Zeiberg: "Kent thinks Crabtree is the smartest guy in the Air Force. In a different system Crabtree would become Chief of Staff of the Air Force."

Zeiberg: "There's a funny story about Crabtree. In the middle of the hottest days of MX, one day Hepfer gets a message from Air Force Personnel which says Lieutenant Colonel Bill Crabtree should be reassigned. 'According to our personnel analysis, we think he should be reassigned as a special assistant to Admiral So-and-so, in the joint Chiefs of Staff.' Hepfer hits the ceiling. He calls his counterpart in Personnel. The guy looks up the record and tells him that Crabtree doesn't have staff experience in the Pentagon! That if he's going to be promoted he needs staff experience! By that point in time the Chairman of the Joint Chiefs, General Jones, was calling Bill by his first name, Harold Brown was calling him by his first name. He was calling Perry by his first name. The guy was really one of the key people in the military-industrial structure."

Michael watched this small man who had drawn such praise for his central role. Michael sat very still. He did not look at Buttimer.

Crabtree glanced at the wall, then at Michael. He tapped on his shoe.

"We came to a general conclusion — that there was not a consistency of data to do a comparison. There had been consistent work done in the past — the Strat X study was done — in the late 'sixties — for example — but in terms of up to date work there was no consistent comparison available. So we undertook to do a comprehensive — comparison — of several concepts. The concepts were the trench — the horizontal shelter — the vertical shelter — and the pool. Those four concepts were compared — in detail — exhaustive detail. Cost — people — nuclear effects — everything we could think of to try to illuminate — the differences between those four concepts."

Michael asked:

"We?"

"That was we, meaning BMO people. BMO and TRW. TRW, in ancient history, space-technology-wise, has been the contractor to do both our — technical analysis and our review work. TRW has a lot of — corporate memory for the ICBM process."

Michael looked sharply at Crabtree. Crabtree glanced at Buttimer. Buttimer licked his lips. Crabtree looked back at Michael. Michael's face was clear of emotion. He wrote in his book: "Corporate memory!" Crabtree unhooked his leg from the chair arm and crossed it over his other knee. He continued:

"In May nineteen seventy-eight General Hepfer recommended — that the vertical shelter be adopted as the basing mode — because it was lowest cost. At that point we started very rapidly — turning our in-house developmental efforts to focus on the vertical shelter — we developed a program to do some early experimental work — at our engineering test bed — on the Nevada test site. We particularly focused on the Transporter-Emplacer — which I guess today is the world's largest vehicle that moves at any speed."

Michael asked carefully:

"Is this the first time you have been involved in an issue that has had such a tremendous political impact?"

Crabtree put his leg back over the chair. He released a short sigh.

"Certainly this large. The A-10 — was to a fairly substantial degree — a political issue. But it was not on the national news each evening like this program has been."

"Do you find the politics interesting or would you just as soon not have to deal with it?"

Crabtree tapped his shoe.

"I'd just as soon not have to deal with it. But I — early in the A-10 work — I discovered that they're inseparable."

Michael nodded crisply. He asked: "Does dealing with the politics require a different personal approach?"

Crabtree glanced at Buttimer, then looked at the floor.

"Communication in the political world is — at least to someone like myself who grew up — in the technical world — is drastically more complex — and more difficult — than communication in the technical world. In the technical world there is a set terminology — that's fairly unambiguous — you're usually dealing with people who know as much or more about — whatever you're talking about — as you do. There is a way to write things down — so if there is any confusion — it's fairly easy to resolve it."

Crabtree looked around the room. There was a trace of his West Virginia background in his voice, a little backwoods ring. He went on.

"None of these things is present in the political arena. One has to find a whole different set of terminology to communicate with — the terminology you use in the technical world is totally inadequate — one frequently sees the terms being misused. The real communication that goes on that's important — in trying to further some perspective — in trying to get somebody to accept a perspective or understand a perspective — are by and large all informal — with all the pitfalls that that has — in terms of whether or not you can really motivate people to understand your perspective — in terms of trying to tell whether they've understood what you're trying to convey or not. The — formal communications — are almost always — adversarial. In the formal communication — you're almost always more concerned with whether or not you're saying something that's going to be taken out of context and used against you than you are in terms of communication."

Michael said politely: "Yes." Crabtree rolled methodically on.

"Sometimes you leave a conversation — and after you leave and think about it for awhile you suddenly realize that —"

Crabtree laughed. Michael sat up. It was a genuine laugh, springing like a trick clown out of the solemn container Crabtree seemed to be. Michael muffled his surprise. He offered a smile. Crabtree went on.

"You suddenly realize that what was going on there — had nothing to do with what you were trying to focus on. And — there is almost never — a second chance. The political process — looks at an issue — and then it moves on."

Michael said coolly:

"Can you think of any specific examples of conversations like that?"

Crabtree chuckled. He looked at the floor. He said: "Not for publication."

"Do you have any way of identifying moments of your own life — in relation to MX or some of these other periods — which were turning points in your life; learning to cope more with the political angle?

Crabtree glanced at Michael. Michael's face was set. Crabtree looked at the wall. Suddenly, as if in response to what he saw in Michael's face, he was humorless again, remote, as if he had rebuilt a barrier that had momentarily slipped.

"No," he said. "No. I'm not very introspective in that sense."

Crabtree shifted, and put his crossed legs out in front of him.

"Folks outside the Air Force — the Department of Defense — may not realize the enormous lack of decisiveness that the Air Force has to deal with in terms of trying to defend the country. The B-1 is just a — classic example. The Air Force had to have itself in a position to produce the B-1 for President Carter. President Carter elected not to do it. The Congress told the Air Force to continue the — test program — the R and D program, and to maintain itself in readiness to produce the B-1. President Reagan came along and said do it. That takes — in my view a level of professionalism you don't find in many other places."

Michael looked at Crabtree. Crabtree looked at the floor. Michael said coolly: "The development of that *professionalism*. You're a pretty good example of it. I'm interested in how that professionalism is developed. Is it that the Air Force attracts people whose nature is to —"

Crabtree broke in.

"I think to a degree that probably is the case. When I left college and when I had to make some major decisions about whether or not I'd stay in the Air Force there was a term — relevancy. People wanted to work on things that were — relevant. Rather than science just for the sake of science or engineering just for the sake of engineering. People wanted to work on something that had some good for mankind."

Michael looked hard at Crabtree. Crabtree was studying his shoes. Crabtree looked up at Michael. Their eyes met. Michael looked down at his book. He wrote: "good for mankind."

"And in terms of the type of work that I have done, I think it does tend to — draw people who are motivated to a degree — they have a lot of other motivations as well, including very selfish personal ones — that are motivated to a degree by a desire to have their work be — be useful."

He paused. Michael wrote in his book, "Useful." Crabtree continued.

"In my own case I have — spent a large part of my career — just trying to understand what it was that — people like myself should be working on — so the Air Force can carry out — its responsibilities."

Michael looked up. Crabtree's eyes were steadily on his face.

Responsibilities!

Norton Air Force Base, 1978. Gentlemen. Gentlemen. The Soviet buildup continues unabated. Gentlemen, it has come to our attention that the Soviet MIRVing of its largest missiles is proceeding rapidly. Gentlemen, we have learned that the Soviet Union now has approximately 5,000 nuclear warheads. Gentlemen, our targeting analysis indicates the Soviets can now strike with lethality at hardened silos. Gentlemen, the Soviet Union now has 500 more strategic offensive launchers than the United States. Gentlemen, the Soviet Union's ballistic missile force now exceeds and surpasses the United States force by a total of 1,464 warheads.

GENTLEMEN: WE HAVE LEARNED AS OF NOVEMBER '77 THAT, TO OUR SURPRISE AND GREAT ALARM, THE SOVIET UNION IS INSTALLING ITS NEW, MORE ACCURATE GUIDANCE CAPABILITY IN ITS EXISTING INTERCONTINENTAL MISSILES!

Michael's time was up. Crabtree stood. Michael reached out and shook his hand. "Thank you," he said. He smiled.

As Michael left Crabtree's office he had to excuse his way through a crowd of men standing in the little anteroom. He was momentarily startled: There was Gary Aubert, from RDA, Aubert of the pins in Russia. There was Alan Schaffer, of TRW, Schaffer of the ficus plant and the white hair. Aubert, his ice-hockey player's face still sharp and cool, grinned at Michael, and studied him with his blue eyes. Then the men filed into Crabtree's office and closed the door.

II

In a familiar motel room, the lights of the trucks flashed on the curtains. Michael walked around the room. In the moving shadows, Crabtree was before him. This little man, this seemingly cold-blooded, corporate-memory missile engineer, with his contortionist's body, his diffident eyes, his jargon. He sat before him, invisible, tangible, shifting his gaze, tapping his shoe, hiding under these layers his inexorable determination, his passion. The flash of lights through the curtains caught Michael's wide eyes, eyes startled — by what? The noise outside the room? His sudden clear empathy for this man? Abruptly, Michael sat down at the little table and wrote Bill Crabtree a letter.

17

I

"A FRIEND of mine told me a story," Big John Toomay said. "This was back East. He sent his teenage son out to get some bread for dinner, and he didn't hear from him for three days. Finally —" A grin developed on Toomay's massive face. "— Finally he got a phone call, and his son said, 'Dad, I'm in the Tijuana jail, but I can explain everything!' " Toomay laughed. "Well, that's the way these systems are. You know, they become — when you finally see the finished thing you could say, 'How could anybody generate such a preposterous thing,' but then when you go through the convolutions of the thought process that led there, in the end you say, 'Oh, yeah, I see now.' "

Michael grinned at Toomay. He settled back in the white chair. Virginia Toomay had brought them each lemonade. Toomay slumped in the couch. When he walked he walked stiffly, but sitting down he still looked like a basketball player. Michael watched him with affection.

The route to the Tijuana jail. Spring of 1977. In the Arizona desert the trenching machines bite into the sand, under the hard eye of Larry Molnar — The *warfighting* capability of the United States is in the people, in the *people!* The people sweat and dig. In Washington Bill Perry calls in Dr. Michael May from the Livermore weapons laboratory, and Seymour Zeiberg and Lew Allen call in Big John Toomay from Air Force Systems Command. Their task, in words never spoken thus, is to kill the trench.

Perry: "When I really started looking seriously at MX designs, in the spring of nineteen seventy-eight, I concluded that the linear trench was a loser. Other people in the nuclear community, and Al Latter was one of them, were coming to the same conclusion. By the spring of nineteen seventy-eight we had dismissed the linear trench. Some people were still thrashing around with it, but from my office's point of view it was a dead duck."

What still has to be done, however, is the formal execution of the bird itself. And here, again, are Seymour Zeiberg and John Toomay.

Zeiberg loves this kind of politics. Work it behind the scenes. Put some group up there to do the unpalatable job that needs to be done. Validate a decision already made: "I worked in the shadows whenever I could. Manipulated the schedule of events to maximize the opportunity. I kept the Hill, ah, informed. I kept certain members of the media informed. I did similar things in the bomber business. I was able to keep the B-1 test program going. I kept things burning in these areas, which were ostensibly contrary to nominal administration beliefs. A very important guideline when you operate in these gray areas is a careful avoidance of asking for directions. Never ask for clarification. You get the wrong answer, see."

Meanwhile, John Toomay approaches the same objective from almost the opposite perspective. He loves the way of a Tiger Team: a fierce application of honesty and knowledge to a problem either to pulverize it or transcend it with an invention — of a new machine or, better yet, of an old truth that has never before been understood. He loves to bring clarity, logic and intellectual courage to a confused picture. "I am a principled guy. I don't consider that necessarily admirable. What I'm trying to do is explain life to myself; that's why I don't believe in ghosts or transcendental things or supernatural things — as soon as you believe in all those the possibilities in the world become infinite, and your ability to control everything becomes zero. So it's much more simple if you can think of the world in terms of a few fundamental things, and that's what I try to do and I try to stick to that. So when some general comes in and says I want this because I like it, that's when I jump up and say NO! goddamnit, we've got to have RATIONALITY! That's really my one objective, is to be that way."

In the desert, a man named Ray Hansen, from Spokane, Washington, is inventing that marvelous machine which eats concrete and excretes tunnel at seventy feet an hour. In Washington Big John Toomay is putting together a group of men that comes to be called T to the fifth: the Toomay Tinker Toy Tiger Team.

Toomay sipped lemonade. The ocean breeze rang the wind chimes. He seemed to have no consciousness of the passage of time. He rolled on. "The idea was that the MX was a Tinker Toy which could be put together piece by piece and every piece would be rational and therefore the end result would be rational." He chuckled. "People were saying your tunnel system, your trench system is just ridiculous. So Zeiberg and others said what we've got to do is start over from ground zero, from scratch, and try to see what system would come out if we just objectively looked at the problem and went after the solution. The problem is the vulnerability of Minuteman and the solution is something which retains Minuteman's previous unique capabilities so it will be a viable member of the Triad. They got a guy like me whom they considered an honest broker, to go out and get this team together and do this analysis, one, to give more credibility to this approach, and two, to act as a motivator to the SPOE to do an honest job of their design of the same thing. So our Tinker Toy Tiger Team, which consisted of a bunch of contractors and military guys all thrown into a group and thrown together to study this thing, went back and tried to do that; we didn't do a perfect job, but we tried.

"I picked the members of the T Five team. In some cases I had to ask a friend to provide a guy that had capability. We had contractors and we had Air Force guys and we had D.O.D. guys; in a sense I had every guy I wanted to have in the Air Force.

"It was a lot of fun because it was just *atavistic*. I went back to my old role of a guy organizing a few guys to get a particular job done. Boy, their energy! They'd been trapped in the bureaucracy and as soon as they got out for this three months they just blossomed. God, they did magnificent work. I sat there cheering them on. That must be something we humans do better than any species; get a group of no less than twelve and no more than fifty, and get something done. That must have been the size of the groups that planned hunts and went out and killed mastodons.

"All we did was work. We were extremely informal. We were in a building at a place called Bailey's Crossroads out in Virginia. There were about twenty of us.

We'd meet in the morning — packed that room — then disperse. There was a Roy Rogers restaurant that was right next to this place, and it was snowing that winter and stuff. At lunch we'd just walk over there through the snow, have a Roy Rogers thing, come back and go to work.

"We had some of the members of another subculture, the guys that have exploded nuclear weapons down at Nevada. We had three or four. They referred to each other — there was a guy, Lieutenant Colonel Lou Montouli, they called him The Wop. There was a colonel they called The Fat Man. My deputy on the study, they called The Red Head. Those guys had a community of their own: They had detonated underground tests, and they'd been exposed to a number of roentgens, a hell of a lot higher than anybody else. They thought of that as a badge. 'You guys are bitching about what's coming out of Three Mile Island, let me tell you what happened to ME. I got this many rems; I haven't noticed anything and that's fifteen years ago.'

"We had this one young lady who was terrifically competent. She could type like the wind and read anybody's handwriting. She was damn good looking and sexy as hell. She had a motorcycle, pictures of herself on the motorcycle, and she was sort of — the social activities of that group centered on her. The young men that were in that group all tried their hand at taking her out and all that stuff, but nothing permanent developed, but she became a sort of mascot of the group."

Toomay paused. Mascots for the MX missile, early spring of 1978: the beautiful girl, the towering general. Duffy: "Toomay's a big, robust, healthy athlete to start with, and he's a very bright man to go with that. He's articulate, physically impressive, and, I think, brave and outspoken. An open guy. Toomay is a man's man kind of guy."

"We did an enormous amount of work in a little time," Toomay said. "The guys would tell me: 'God, this is terrific!' I had colonels, lieutenant colonels; they were all excited because they'd just been doing staff work until it was just grinding them down into nothing. Lots of young officers who were embedded in the bowels of, say, the controller's office. Suddenly they were freed to price this big weapons system. Civil engineering office systems command, there's a young Army guy who worked there, doing some mundane thing, and we put him to work finding out how much it cost to build roads, railroads, underground roads; he was just in ecstacy. We'd have meetings and the MX SPOE guy'd say, Yeah, roads cost this much, and this guy'd jump up and say, I don't know how you got that: I did these road designs based on this manual —

"We really cleared the air. The SPOE guys said to themselves, Hey, we can't get away with anything. We have to show everything. John Hepfer will probably tell you he thought that the T Five group was the worst damn thing the Air Force ever did because all they did was come out there and muck up his operation." Toomay laughed. "Yet he was a good friend of mine.

"Crabtree was the guy who was redesigning the system as we were yelling and screaming at him. He started working then seven days a week for untold hours every day, and has been working like that ever since.

"We found that the steps by which they'd arrived at their conclusions had a bunch of *logic* flaws in it because they had been done over a long period of years in no coherent way. We went back and tried to do those steps very *logically*. If you want to protect something, you can do four things — you can maneuver quickly so the guy cannot spear you — I remember the chart; we had a big guy with a big spear — or you can put a big shield on your arm and protect yourself that way, or you can . . ."

"We went through the whole *logic* chain, the legal brief to get where we got: You can't have the missile moving because if you have it moving it doesn't have enough hardness and you've got to use this big portion of the United States, and so on. It's stupid to have it in a tunnel because roads on the surface cost less. It's stupid to have it in a tunnel because once he finds out which half of the tunnel you're in the system's defeated. And you can't move back and forth in the tunnel without his detecting — that kind of thing. We gradually, with a purely *logical* approach, drove right down to the multiple protective shelter with a vertical silo."

"We dug out these things, these *technical verities. Technical truths.* What happens is, something may be true but if the community doesn't understand it it doesn't do any good for you to bring it up. So it's always important for the community to have diffused through it the knowledge that's been generated. As soon as that happens and you learn to hear it and live with it, then the group, the community, all the guys that have anything to do with it eventually become accustomed to it. Somebody has to tell you that that's the case, then you think about it, and after that, if enough people know that, it gets to be *wisdom.*"

Wisdom: Bill Perry: "By the fall of 'seventy-eight, it was quite clear the Air Force establishment had swung over to the multiple protective shelter system. The corporate wisdom within the Air Force was that the MPS was the way to go."

Toomay had stopped. Then he said, "Say, you want to stay to dinner?"

"Well —"

Virginia Toomay's voice came from the kitchen. "Guess what," she said. Her husband said, "What?"

"It's tacos again."

Michael smiled and relaxed.

"Sure," he said.

II

Michael sat in a street-level cafe in the San Francisco financial district, trying to eat at a piece of carrot cake he didn't want. Bill Perry was late. His office had told Michael to come back in an hour. Michael picked at the cake. He watched the proprietor of the place, who wore his name stitched to the front of his shirt like an auto mechanic, cleaning up in the afternoon doldrums. Michael saw explosions. Pieces of wood, concrete, dust, flying glass, darkness. The man, who eyes were sardonic and whose mustache was soulful, put great thick plates away in a cabinet. Michael saw him vanishing in a gust of impossible heat. The man offered him coffee. Michael saw this place collapsing under millions of tons of unbonded brick. It does not happen as it did in the movies; in slow motion, brick upon brick spinning, falling, shattering — it happens in a second: Pour the coffee — slam! Gone! He looked at the man's face. The man looked at him in boredom. Michael saw photographs of the man with his children taped to the wall. Michael almost said: Don't touch me. I carry illness; I carry evil news.

That had been the role of Bill Perry. Michael had a news clipping in a file folder he carried with him. He sat at a table in the doomed cafe, and read. *Newsday*: "First in a series:

Washington — As William Perry walked into his Pentagon office that morning in early November, 1977, the intelligence liaison officer already was waiting for him. And the officer seemed excited.

As the deputy secretary of defense for research and engineering — the third ranking official in the Pentagon — Perry received a regular Monday morning intelligence briefing. But, as this briefing began, Perry realized it would not be routine. There was new information about a major advance in Soviet missile systems.

Gentlemen! Data gathered in recent days indicates the possibility that the Soviet Union has, in firings of its SS 17 and SS 18 missiles, demonstrated an ability and capacity to achieve approximately a CEP of 0.14 nautical miles, far surpassing that expected in the current generation of Soviet ICBMs!

In his folder Michael had filed others' remarks about Bill Perry.

Albert Latter: "Both Bill Perry and Harold Brown are very bright, very substantial people. Their standards of excellence are high."

A House of Representatives staffer, opposed to MX: "Bill Perry is poised, able; a very caring human being. His heart is good."

Dr. Sidney Drell, a physicist opposed to MX: "I have the highest personal regard for Bill Perry. Bill has been quite honest."

John Hepfer: "Perry was one of the finest managers and technical people they ever had in that top job. Perry was very precise and methodical in his process. He knew what had to be done."

There were other clippings in Michael's file. When a writer for *Harper's* magazine visited Perry in Washington on a mission to attack MX, the writer noticed Japanese silkprints hanging on Perry's Pentagon wall. "He is fine-mannered and not given to converse in military-scientific jargon," the writer later remarked, "a trait that has won him admiration and trust on Capitol Hill. He offers the visitor tea. On the low table between sofas far from his desk is a cut-glass vase holding a single, perfect pink rosebud."

And a *Science* magazine writer, asking Perry for an evaluation of the Reagan administration's debate over what to do with the MX, wrote: " 'The worst of all possible solutions,' according to Perry, would be to place the new MX missiles in the old silos that now hold Minuteman III missiles. . . . It would probably force the United States to adopt a policy of 'launch on warning' . . . Perry says he cannot even discuss this option 'without breaking into tears.' "

The hour had passed. Michael paid for the carrot cake and went upstairs. The offices of Hambrecht and Quist looked accommodating to a man who liked tea and roses: tall glass doors, dark paneling. Perry, out of breath, swept into his office an hour and a half late. He invited Michael into his office. He looked mildly sad. People often greeted Michael with both weariness and suspicion; in this case, on this late and windy afternoon, Perry seemed to be working hard to conceal both evident emotions on Michael's behalf. He smiled; he was gracious; he listened to the inevitable garbled question without impatience.

"Do you remember the moment — the firing of the Soviet missiles; the new accuracy — do you recall your reaction?"

Perry was a slender, dark-haired man, a mathematician by Ph.D. He was urbane and soft of voice. He wore a dark suit, a white shirt and a dark tie. His hands were thin. His dark eyes seemed contemplative, and, as if what they contemplated was unwelcome, unhappily wise. The Gentleman of Science.

He nodded slowly, and offered a smile of enormous sadness.

"I can vividly recall the moment I heard about the first firing," he said. "I instantly leaped to the conclusion that turned out ultimately to be the correct conclusion. That was an intuitive leap forward. It was quite clear that that firing gave considerable urgency to the task that even in this office we had not felt before that. Because that firing, particularly as we begin to digest its whole significance — that firing indicated that the Soviets were not going to wait for their next generation ICBM to get their accuracy."

Perry's two-page biography, also nestled in Michael's file, spoke briefly about his ten-year career as founder and president of ESL, Incorporated. "Besides his management duties, he engaged in analysis of missile systems and the design of electronic reconnaissance systems." An Army Civilian Service Medal cited him for "the development of systems for the collection of vitally important intelligence through the use of advanced electronics." Michael had recently talked to another MX opponent. His notes on that interview were in his book:

"How did he *know* the missiles were accurate?" the man, Christopher Payne, at the American Federation of Scientists, said. "With the methodology that *Bill Perry* himself had designed for ascertaining Soviet accuracy. ESL built strategic and tactical reconnaissance systems for the CIA, the Army. So it was *Perry* who sounded the alarm. I have a slight *problem* with having an important official verifying his own verification systems."

Unperturbed at Michael's less vivid translation of those remarks, Perry was vague.

"ESL's specialty was the design of reconnaissance equipment," he said, with a gentle verbosity. "But I'm really not allowed to discuss what applications were made of them. It is true we did analyses of data, and I personally did some of that analysis of data at ESL. It is probably fair to say that my background made me appreciate immediately the significance of the data I had seen and also gave me an opportunity to make a personal critique of the analyses which were presented to me by the intelligence community."

Perry tilted his head slightly, with an air of graceful sorrow.

"It was quite clear that that firing gave considerable urgency to the task that we had not felt before that. Because that firing, particularly as we began to digest its whole significance, indicated that the Soviets were not going to wait for their next generation ICBM to get their accuracy. They were going to make a modification to the existing ICBMs. My judgment then was that when this new ICBM guidance system was finally deployed in quantity, the Soviets would have a counter-ICBM capability."

Bill Perry's office in this sedate old San Francisco building was tiny. His desk hugged the wall. In here Perry looked frail. His suit looked a fraction of a size too large, just enough to make him look gaunt. Michael watched him. In the summer of 1978 Perry and Harold Brown took that message into the White House to persuade a President who had once called for huge nuclear arms reductions.

Summer, 1978. A light step, a quiet voice. Bad news carried with reluctance, with sorrow. Mr. President, alas. Mr. President, beware. Mr. President, the Soviets. Ah, the Soviets. An MX critic who had worked for the Air Force: "Bill Perry — he and Harold Brown are most impressive figures. They are very sincere, very thoughtful, are nonpartisan, have enormously high IQs, have a long background in the business. When Bill Perry or Harold Brown would meet with . . . especially with small groups, I think they were just overwhelming."

Gentlemen: Alas, beware! Mr. President, alas, beware! Bill Crabtree: "Dr. Perry brought an enormous amount of leadership — to the task at hand — he convinced a lot of people — to do things — that emotionally they didn't want to do."

This summer that one message is Bill Perry's task. All summer Bill Perry works, a slender man with a mathematical mind, a man with thin pale hands, a long face, and a plain dark suit, a priest of a solemn order. He works, passing his message along, into the highest mind. Perry: "The President was very cool and analytical in meetings." Perry is very cool, very analytical. The Gentleman of Science. Albert Latter will never be in this position. Latter is restless, wild, intense, flinging his eyes and his mind about like a torch, igniting all within reach. Perry is so quiet, so apologetic; gently, firmly moving in the direction in which he chooses to go. Is it odd that the Boys behind the Bomb, by circumstance, sent this man of gentle aspect forward with their most intimate, brutal message? Mr. President, it has come to our attention that the Soviet Union is now capable — Mr. President, we have learned that the Soviet Union will soon have the power — This, the most precious and fragile rite of belonging to the greatest fraternity of power in the world, the passing along of the burden of the knowledge of the danger, so the man who had wanted to thrust these armaments away could instead become stiffened and upright and strong, has been entrusted to this pale gentleman. And it is a masterpiece. Who could be more subtle? Who more tough?

Victor Utgoff, a staff member of the National Security Council:

"Perry's persuasiveness is a function of his personality. The way he thinks. Bill Perry is persuasive on everything."

Michael: "So the fact that the President gave weight to what he was saying had nothing to do with the moment in time. It was because the President trusted him?"

Utgoff: "Yes. When you're listening to Bill Perry you can tell. You're listening to a mind that's very careful in making sure he's covered all the bases. You can hear him sort of ticking off all the possibilities, and it's clear that he's looked at them all. It's the man. I want to be credible when I talk about Bill Perry's ability to convince people of things. But I will tell you, it is truly impressive. He knows his shit."

In his office at Hambrecht and Quist Perry said softly:

"Once we explained the significance of those intelligence data to the President, he accepted that the problem was real."

Perry concluded the interview with wonderfully regretful politeness. Michael shook Perry's hand and departed. Somewhere in the building an elevator was sinking or wind on the street sucked at an open window; under that pressure the big doors of Hambrecht and Quist moaned. Michael pushed them open, relieved them, and stepped out into the silence of regained equilibrium, into a hallway of dark oak-paneled walls. The summer of 1978. The sudden change. Hepfer: "You could sense everything being negative and all at once there's urgency to find some kind of solution even if it's a little crazy." Utgoff: "You could feel the acceleration after that. All of a sudden things were moving."

Michael walked down to the street. Dr. William Perry's face remained before him, and he heard his unassuming voice: Alas, beware.

III

Michael caught the BART subway at the Montgomery Street station. The train sank into the earth and rumbled through the tunnel under San Francisco Bay. Michael looked out the window at rushing darkness.

The tunnel. Late summer, 1978, on the desert east of Yuma, Arizona. It is time for the test. The test is designed to help BMO learn whether the Strongback can indeed shove the missile out through that massive shell without smashing the missile itself. Two Strongbacks have been shipped to Arizona and assembled. One had been built by the Martin Marietta Company of Denver and the other by the Boeing Company of Seattle. The Strongbacks are placed inside a short separate section of trench.

Heat, mirage, sweat, a few VIPs in a little grandstand. Some television and press photographers. Scaffolding bearing the official cameras. A hot wind. Temperature in the shade is estimated at 110 degrees. A group of Boy Scouts from Yuma is selling soft drinks. Larry Molnar is here. (It's in the PEOPLE.) John Hepfer is here; people salute his vast grin. Hepfer will recall the day forever because the air conditioner on his car has failed. Sweat. Restlessness. Above the hidden Strongback is a large pad of pale earth which has been crisscrossed with chalk lines so the cameras will record precisely where the Strongback emerges.

At a signal, the cameras whirr. The ground heaves, buckles, splits, spits dust, rises — all in slow and steady motion. Clouds of pale dust rise and blow south; the crowd stirs. Slabs of earth slowly appear, then rise and fall away like chunks of light rock floating upward in lava. Out of the heart of the earth there rises inexorably an enormous yellow cylinder, shedding debris as it groans upward to tilt on its gleaming neck like a beast out of the Mesozoic Era awakened by a wizard.

The vast yellow creature reaches its height, squirting water from its hydraulic joints, standing clean and free in a swirl of dust. Now, at the moment when it should release its millennial pent-up roar, all noise and action ceases. There is a small chirping of applause from the crowd, and grins on the faces of some Air Force men in the front row. Voices are heard in the silent flow of the wind. Hot damn! By God, the buried trench works!

John Toomay:

"I wasn't there. I saw the movies, though. I saw them again and again."

Michael: "What did it look like?"

Toomay grins. He hesitates a moment, then, typically, bursts out with the unescapable metaphor:

"It looked like a GIANT erection."

18

I

"WHERE," Michael asked, "is the stress in this business? For someone like Zeiberg, Crabtree, or Toomay. The stress that is applied to one's character — through which one grows as a person?"

Dr. Michael May looked gently amused. Michael sat facing him in an odd little office somewhere on the edges of the Lawrence Livermore weapons laboratory. Michael had been escorted through a tall gate, past rustling eucalyptus trees, past an auditorium where a group of men had been standing taking a coffee break — their name tags all said PYROTECHNICS — and upstairs to an area prominently labeled "Unclassified." There, in this long, narrow office, he had met Michael May.

May had been the chairman of the committee appointed by the Carter administration to study the fate of the MX missile at the same time as the T to the fifth team. That was why Michael was there, although they had only talked briefly of history when Michael changed the subject.

May was a very small, thin man. His face was as intricately wrinkled as the great Southwest desert seen from space. His eyes were brown, and kind. In one of his windows there was the black model of a web and a spider. Both he and Michael had tape recorders running in the middle of the table.

After Michael's question, there was a long silence. At last May said kindly:

"There is always a learning experience. Some might tell you that their learning experience is learning to doubt. One thing that everybody learns is that there is not going to be any final answer to anything. You are engaged in a long evolutionary process. Once you say that you're going to provide a nuclear weapons system which can survive attack, and deny any advantage to the attacker, there ain't going to be any one-for-all answer, and furthermore the answer is not going to be dictated just by what looks best on paper.

"MPS is a case in point. If you just take a purely theoretical approach to that, you say look, there are three million square miles in the United States. You can make missiles much safer than dynamite. On paper you can have an invulnerable system on land just by building twenty thousand simple shelters and shuffling it among those things. But that's a system that's politically impossible. But you really do have to take your task as devising weapons systems that are going to defend these actual United States, with real people in 'em, and with a real Congress voting real money.

"Now, everybody learns that. And, if you come at it from a scientific background, that's the part you don't take for granted, that stresses you. And if you come at it from a different background the politics is probably the part you would take for granted, and what stresses you there is to understand that two and two never equals five no matter how much you might wish."

"Once," Michael said, "there was a certain amount of
— anguish on whether or not atomic weapons were something one should bring into civilization. Out of that came
various people who have different viewpoints now. Among
the people who are still involved in nuclear weapons now
there seems to be a consensus that . . . that there is not
any real personal sense that they're doing something that
could have a negative impact on civilization." He paused.
"I mean — is that true?"

May spoke with such kindness that it seemed he must
have seen signs of discomfort in Michael's face:

"I don't know. I don't — think so. I can characterize
my own attitude, and I don't know how general it is. Just
about anybody who works on nuclear weapons has
thought about the implication — that I know. I don't
know anybody who shrugs it off. My own — I think it
is increasingly clear, whether it was through World War
II or some other way, nuclear weapons and other destruc-
tive mechanisms were going to be introduced into civilization. They are part of what we learn how to do. We
learned how to prolong life, and although everybody
doesn't eat well, everybody could eat well; there's plenty
of resources for that. Everybody can live materially well
without running out of anything. We learn to solve these
problems and we also learn how to destroy each other
with greater facility."

May's French accent caressed his words. His seamed
face was steady, his slight body still.

"Now, I think that the more you realize that it's an
inevitable concomitant of the rest of our human civilization, that you learn to do all these things, the more you
feel that the problem that has to be solved is a human
problem. The problem is how do human groups learn

to get along on a worldwide scale without wars. You never completely eliminate violence and aggression, not even within a society, but you sure keep it down to where you don't have to have your own little private army to go from here to Washington. We don't know how to do that internationally. Nuclear weapons are a dramatic means to make that problem real, but if it hadn't been them, it would have been something else. Trying to disarm doesn't deal with the problem because no one's going to forget how to build them. Whether we or the other nations can work out methods for resolving international conflict or not, I think that will help determine what happens to our civilization. I've got little doubt that in the long term the human race will learn how to do it, but whether the United States and the Soviet Union will be the ones who lead the human race, I don't know. It is well beyond simpleminded solutions. Feeling guilty about it doesn't help, and trying to call for the abolition doesn't help, because all that counts is the idea and the knowledge and that's not going to go away."

There was silence in the long room. Michael looked at May's quiet face.

"So," Michael said, finally. "The problem is war, not nuclear weapons."

May did not smile. His head gently inclined.

"The problem is war. Exactly."

There seemed to be little else to say. Soon Michael got up to leave. He stood up and shook hands with May. Through the window he could see acres of level buildings. Somewhere out there in this laboratory men were completing the design of a new warhead for the MX, which would increase the power of each RV from 335 kilotons to about 500. Michael started to walk to the door.

"And what stresses you?" May said kindly.

"Oh!" Michael stopped, as if hit. "What stresses me? Well . . ." He turned and looked back at the little man in the doorway, whose face was so like the weathered face of the desert, whose brown eyes regarded him with such compassion.

"My stress —" He said it slowly, as if it was hard to get out. "I guess . . . is trying not to let my own angers and fears — prevent me from seeing what is really there."

May looked at him for a moment, and smiled. He said, almost formally:

"I am glad that is your stress."

II

Bells! The chimes rang and rang: carillon bells singing across the campus of the University of California at Berkeley. Michael looked out the window at a vast stone wall. It was the wall of the campanile tower. Its edge was severely true.

Michael was sitting beside a small desk in a corner office listening to Dr. Charles Townes. Townes was the University Professor of Physics at the university. He was also one of the inventors of the laser. He was awarded the Nobel Prize for physics in 1964. Michael sat and listened quietly. He looked regularly down at his notebook, as if to occupy his mind.

Townes had been the chairman of the committee formed by the Reagan administration to study the future of the MX. That was, ostensibly, why Michael was there. He had asked how Townes had reacted to being named to the position.

"Controversy doesn't particularly, ah, upset me," Townes said. "One has to accept a certain amount of controversy, and the real question before I accepted this job was to what extent could I do something good. I commented to some of my friends that most of the other people who might do this were connected with the situation either through their business connections, in having a substantial financial stake, or through various organizations in the way they had committed themselves in the past. And I had, ah, relatively little to lose; the only thing I had to lose was time and my, ah, friends, and my reputation. So why shouldn't I do it?"

Michael laughed briefly. He had brought more of his news clippings. As the summer had continued they had accumulated in drifts. He had arranged them in a rough chronology: "The Dilemma of the MX Missile." "3 to 1, House Backs MX Mobile Missile; Site Choice Deferred." "Airborne MX Reported Eyed by Weinberger." "The Case Against the Flying MX." "Reagan Is Seen Leaning to Reviving B-1, Ending Desert-Based Mobile MX System." "REAGAN TO RECOMMEND AN AIR-LAUNCHED MX." "Some on MX Panel Favor Air System." "Single Silos Seen Favored by U.S. Panel." "Tower Says Air-Based MX Could Crash on the Hill." "Congress Held Likely to Reject Airborne Missiles." "Advisers Said to Prefer Land-Based MX Plan." "Airborne MX Version Finds Foes." "Air-Based Missile May Be Ruled Out." "Reagan Is Believed To Back Reduced Land-Based System." "Reagan May Turn To MX Missile Plan Based on Carter's; Airborne Idea Losing Out." He asked Townes about the flurry of conflicting stories.

"The press made all sorts of trial balloons itself," Townes said mildly. "It would build something as being the decision and then have lots of arguments against it, and it was really ridiculous. All sorts of supposedly confidential things coming from high-level people, that frequently had no basis."

Michael asked: "The recommendations you made are still . . . classified?"

"Those are still confidential, yes."

Michael looked down at his notebook. He looked up, an odd look of strain in his eyes. He rattled off a question. "Did you talk to Seymour Zeiberg?"

"Yes, he met with us. He's very strongly for the MPS system. He feels it's the right thing; he's worked on it hard and has, ah, pushed it for some time."

Michael started to say "The question —" He stopped, and instead asked:

"Bill Crabtree?"

"Yeah. Yeah."

"Did you talk with him early or late in your proceedings?"

Townes chuckled. "Oh, all through. All through. We were always calling on him, and he worked like a dog. He's a person who's technically very knowledgeable, very bright, very conscientious, very careful to try to be accurate. It was detectable what his own preferences were, but on the other hand he was not, ah, overtly selling things."

Michael rushed on. "How — How would you compare him to Albert Latter?"

Townes smiled again. He was a tall man with just a slight touch of the frailty of age. His smile was genuine but withdrawn, as if the question brought more to mind than he wished to describe.

"Well. Al Latter's much more of a salesman. Al Latter is also very bright and very interesting, but he has also characteristically been a sort of broad-scale idea person." Townes went on dryly. "Latter brought up a number of different kinds of things that he thought were good and commented on things he thought were bad. Crabtree never tried to tell us whether we should do a system of this kind or that. I'm sure he had his views, but that was not his function."

There was a pause. Michael glanced out the window. The campanile tower was made of huge blocks. They were fitted with precision. The blocks were gray and smooth. The tower looked impossible to climb. Michael took a breath.

"The question is one of trust," he said abruptly. Townes's surprise at the non sequitur was concealed. Michael plunged on.

"I guess what I'm getting at is . . ." He paused, looked fiercely down at his notebook, and continued:

"From my point of view, from the point of view of the person on the outside who has not been involved in this community of people, who sees this impenetrable world of the design and development of nuclear weapons, and starts hearing about counterforce and throw weight and levels of acceptable death, there seem to be two ways of reacting to this, other than simple despair." Michael swallowed, and went on. "One is cynicism. You can despise the people, or ridicule them, or think of them as fools." Townes was watching him curiously. Michael went on. "As if by doing that you could somehow feel, if less safe, then more righteous. Or as if you could somehow hope that better people would make better decisions."

Michael took a breath. Townes did not choose to interrupt.

"I guess," Michael said. He looked out the window at the forbidding stone. "I guess . . . I've kind of rejected that approach. So the only other choice this person on the outside has, since he can't know the details of the operation of this community — the other choice, besides despair, is trust."

He rushed on now, to the end:

"What I want to know is why should we trust them?"

Townes laughed gently.

"Why should you trust them?" He looked out the window.

Michael said: "And that's why I guess I'm trying to ask —"

"Yeah," Townes said gently. "Yeah. Well I'm not sure it's possible to live a completely unperturbed life in these times. I think one has to be concerned. Now, how does a person develop confidence that we're doing something sensible? It has to be based on their judgment of the people who are in control. Do you know these people?"

"I've met them." He cleared his throat.

"OK," Townes said cautiously. "Then you can judge the personality to some extent. Well, you know, these people are all different, and they have their own personal style of doing things. They also have a dedication to, ah, defense, and peace. They feel they are doing something that's very important to the United States and the world. Take Al Latter, for example. He's worked this kind of thing for a long time. He's clearly dedicated, trying to help out. Crabtree is also a very dedicated person, and I would say, more self-effacing than most. He just wants to do a good job."

Michael looked hard at Townes. Townes was shading his eyes to look out the window, vaguely like the statue of an explorer. Whatever it was he saw, it was not compelling. Michael, his hands relaxed on his book, asked quietly, a little sadly:

"Is there a shared characteristic that these people have, in dealing with these . . . these uncommon weapons?"

There was silence.

Michael said, "Guys as different as Latter and Crabtree; is there something that they share?"

Townes said slowly, "All of them are concerned about

nuclear war. Mostly, these people are very, very concerned about it, working very hard to do things that they would believe would avoid that kind of danger. That common theme leads them in many cases to building weapons."

Townes smiled at Michael. Michael managed to smile back. Townes said:

"Do you see any common theme?"

Michael laughed, then stopped. It had seemed a harsh sound. He said:

"I suppose I am beginning."

He promised Townes a transcript of the tape. He walked away, down a hall past a sign on a door — CAUTION, LASER. 100,000 NW — downstairs, past a hall in which every door was guarded by a huge red fire extinguisher, and out into the cool gray air of Berkeley. He passed the base of the companile tower and looked up at the immense obelisk of stone.

19

I

MICHAEL stood on the desert. A small hawk rose from
mesquite bushes and wheeled away. A ridge of mountains lay to the west like a mound of blackened bones.
He was in an enormous valley between bare mountain
ranges, a valley that sloped gently away from him west-
ward to a dry wash, and then sloped gently up again to
another heap of bones. Far to the north a single mountain rose like a blue pyramid above the sand and scrub.
The earth was just brushed with life; low bushes, cactus,
mesquite; the blue sky was just brushed with cloud. It
was almost silent: at long but regular intervals — two
minutes? five minutes? — the sound of a very distant explosion swept across the desert like a momentary tremble
of the air.

Michael walked down the graded slope that led into
the earth. The pale brown sides of the ramp were eroded
in parallel gullies; tiny purple spiky plants sprouted between them. As he walked deeper, the mountain range,
the desert, the expanse of view were all cut off, and all
that remained was the sky and the eroded earth walls.

There were broken sheets of plywood in the bottom, a few nests of wire, and a signpost that had been pulled out of the ground and flung in here. Michael turned the sign over. It said: "DANGER: Hazard Area."

The sound of a far-away blast ran through the air. The little hawk cried, over love or loneliness or a mouse. KAAIIEEE! At the bottom of the ramp was the huge mouth of a round concrete tunnel. It was about twelve feet in diameter; recently the opening had been closed to view by a plywood covering. Now all that remained was a framework of timbers that had supported that wall, and a plywood door. Blocking the way through a nonexistent barrier, the door looked foolish, a guard protecting the hole where a vault had been. Michael stepped around the door between timbers and entered the tunnel.

The floor was tiled with mud. Rain and earth had washed down from above and had come to rest there, drying out in the summer. The desert heat must have baked the inside of that tunnel. The tiles were hard. They clattered and rang underfoot. Michael picked up a piece and broke it against the concrete wall. The shards tinkled, like chimes in the wind. He walked along the loose floor, deeper into the tunnel. The air became cool and still.

The tunnel was straight. Ahead of him a line of light reflected from the opening behind him ran gleaming along the roof, then disappeared. But far out in the distance a tiny round ball of light floated in darkness. That short gleam overhead offered no perspective. The light could have been a hundred yards from him, or a mile. Silhouetted in the distant ball of light was a tall shape. For a moment the world spun, and perspective turned on end, and it seemed as if Michael was standing on the wall of a well, looking down at his reflection in the pool at the bottom. He touched the concrete and regained his balance. It was the other end of the tunnel.

But if that was the other end of the tunnel then it was amazingly distant. If that was the other end of the tunnel, then there was a man down there, standing at the entrance, looking in.

Michael spoke. The tunnel took his voice and seemed to carry it away, sending back reports of its progress. As the sound traveled its pitch rose, until it disappeared in a metallic wail. He sent his voice again. Again the tunnel cried with it, then again became silent. He walked in farther. The ball of light at the other end did not enlarge; nor did the figure at its center move. He walked a few more steps forward; his footsteps rang. He looked behind him. The entrance seemed vast, a wall of sunshine, bright earth, sky, calling him back. The plywood door in the wall could be mistaken, at a distance, for the silhouette of a man. He laughed, and the tunnel sang. But he did not walk deeper into its resonant darkness.

He turned and left, clattering back into the sunshine on the brown tiles. Outside, the man with him said: "It's very peaceful here today."

There was another distant boom. "What is that I keep hearing?" Michael asked.

"Oh," the man said, "it's just the gunnery range."

Michael had come to see the remains of the MX trench test site with a man named John Heinsius, an engineer with the Parsons Company. The Parsons Company had the last Air Force contract: to clean up the land and put the trench away.

Michael and Heinsius got back into the Parsons' car Heinsius had brought. There was nothing left of the trench but a long pale brown swath that ran up the gentle slope of the hillside toward the mountains. Everything else was gone. Heinsius drove a mile and stopped. The two men got out of their car and walked over a large, flat, unvegetated stretch of gravelly desert where there had once been a trailer village, buildings, concrete foundations, power lines, life. It was swept clean. It looked as if it had all been blown away, or vaporized.

They drove up the long slope. On another broad stretch of gravel three men were working. There was a dump truck, a front end loader, and a backhoe. The truck and the backhoe were parked, and the loader was moving broken concrete. The vehicles looked strangely gargantuan on the plain, towering above scattered vegetation. There were clumps of balled wire lying on the ground among rabbitbrush plants of similar size, and in one clear area a very small fire burned with an angry orange flame.

"There were a lot of concrete foundations here," Heinsius said. They were almost all gone. The single exception was a small rectangular hunk of reinforced concrete that two men were trying to break apart with a jackhammer. The structure was less than eight feet square, and the concrete walls were at least a foot thick. The concrete was of such high quality that the jackhammer rattled against it with all the apparent effectiveness of a woodpecker trying to penetrate stovepipe. Neither the two men nor Heinsius had any idea how this massive little foundation had been used. Heinsius and Michael got back in the car and were soon out of reach of the clattering noise, which, as if it found no support in the dry air, vanished straight upward.

Heinsius drove the car up a wide stretch of graded earth, sending up a faint plume of dust. Once Heinsius had to negotiate a rut that spring rains had cut in the gently banked earth. "See," he said, mildly — everything about him was mild — "it is already eroding. If you don't keep it up, it goes very quickly."

The two men got out of the car at the end of one section of the trench. The place was incredibly silent. Michael could hear the sound of a truck on the highway murmuring along several miles to the north, and once there was a boom from the gunnery range. There was no sense of direction to the noise. Here there was an opening in the top of the trench — one of the gaps in its ceiling called a SALT Port.

The opening looked like a sinkhole in the earth. It went down to the rectangular concrete lip. Inside the exposed section of tunnel there was a round plywood cylinder, painted white. It was a model of the missile. It had warped in the weather.

"I guess they used that to see if you could detect it from the satellites," Heinsius said. Michael involuntarily looked up at the untroubled sky.

At the far end of the trench — the west end, close under the knobs of the Mohawk Mountains — the Air Force had dressed up the trench opening with a false retaining wall and a wooden model of the steel door, like the round door of an enormous safe, that would have locked the missile in. The concrete veneer of the retaining wall was cracked, showing the outlines of the sheets of plywood underneath it. The mockup of the door hung open on its enormous hinges. The inside of the trench here was, oddly, lined with metal foil. Michael looked inside and saw again the floating ball of light with the figure standing at its base. The figure might as well be him, hearing his fear get away from him and rise in those strange acoustics, to a metallic shriek.

"The contractor has to walk the whole trench," Heinsius said, "to make sure there is no life in it, before he seals it up." There were coyotes in this desert, and illegal Mexican aliens, too; once the ends of the tubes were shut by the 3/8-inch steel plate the contract called for, there would be no escape. The trench would be buried entirely, and would lie beneath the desert floor, forgotten, until one day, those wandering streams cut away the earth to show a strange white hump in the desert.

"It is planned for a fifty-year life," Heinsius said. "Even the paint on the steel is a fifty-year paint."

In the next two weeks the contractors would attach the doors, fill in the ramps with earth and debris, and blade them over, hiding the handiwork of John Hepfer, Bill Crabtree, Larry Molnar, Ray Hansen, and all the others. And when those rains and floods had finally revealed this thing to the future, what would those different people of that time expect to discover in this preserved capsule of twentieth-century space? What would be their disappointment — or their understanding — when all that they found were mud tiles and a plywood missile?

Heinsius and Michael drove slowly down the long, wide road away from the trench, dust rising behind them. They said little. Heinsius negotiated two gullies. (It goes very quickly.) Heinsius pointed north. There, he said, on a nearby section of the long sweep of valley, was the rectangular swath of soil where the short section of trench had been built for the testing of the Strongback. That was where the cameras and the small crowd had gathered that day in the late summer of 1978. It had already been closed and covered, and already the mesquite and the rabbitbrush were encroaching on its outlines.

Michael looked across a couple of miles of desert at that odd space. Within that concrete tunnel, Larry Molnar had told him, was interred the skeletal remains of the two Strongbacks, huge steel bones, stripped of hydraulics, gas generators, and fluids, resembling more than ever the great fossils with whom they were now doomed to share the uncounted years.

Heinsius drove past the last foundation. It was still intact. The two men with the jackhammer were resting. Heinsius drove on, and the figures, the foundation, and the parked backhoe became smaller and smaller under the enormous desert sky. From the great flat where the trailers had been they looked like crows, or ravens, perched near a dead yellow tree, beside a lump of bones. At last they dwindled to insignificance along with everything else on the long sweep of the skyline.

Finally Heinsius remarked, mildly:

"It was a waste of money, you know."

Michael didn't answer.

THREE

The MX Mafia

20

GRAND TOTAL STRATEGIC OFFENSIVE WEAPONS

| United States: | 9200 |
| Soviet Union: | 5875 |

I

THE two implacable young officers of the United States Strategic Air Command sat on either side of Michael. It looked like an interrogation.

Michael sat at the head of a long table. The man on his left was a polished young lieutenant colonel, with gold-rimmed glasses, who looked at Michael with an aggressively honest countenance. His name was Claude Mitchell. The man on Michael's left was equally well uniformed. He wore his identification badge around his neck on a chain. He was a major. His name was Jim Zorn. His eyes were shadowed with introspection. Both men leaned forward, watching Michael.

The interrogation, of course, was running backwards.

"Does SAC prefer the largest possible missile?" Michael asked.

"Absolutely." Zorn.

"It really boils down to the issue of capability." Mitchell. "There's just so much you can do with a smaller missile. It doesn't have the payload capability. The ninety-two-inch diameter missile has the capability we need to do the job that we say needs to be done."

Michael smiled. Mitchell continued.

"What you are really talking about is the combination of nuclear yield and the accuracy. If you start with a weapon which is OK for city-busting, let's say. Soft Area Targets is the general term that we're looking at there. You don't have to have a weapon that's very accurate. The nuclear yield and the ranges that we're talking about, you get that thing fairly close and it does the job. Take that same nuclear yield and try to destroy a hardened point target with it, like an ICBM silo, or like a command control bunker." Mitchell glanced at Michael to make sure he followed. Michael nodded. Mitchell went on.

"In that range, then, the accuracy becomes a lot more important than the yield does. The accuracy is the number that really plays more strongly in the ranges of nuclear yield that we're talking about. The missiles that we now have, if they survive, with the yield and accuracy that we now have are not that effective at all on that kind of increasingly hard target that we've been presented with."

Zorn added, his voice surprisingly soft: "The hard targets are the difficult ones to destroy, and if you're going to buy one missile, you have to size it to do the most difficult job."

Mitchell continued crisply. "Frequently, more often than not, that target also carries a premium for response time. It has a perishable value. It's valuable only if you can get there very quickly, when you decide you need to do that."

Michael asked:

"How does the MX today compare with what SAC asked for back in nineteen seventy-two?"

Zorn and Mitchell looked across the table at each other.

"I think it compares fairly well," Zorn said. "What can you say? That's what we told Air Force Systems Command we wanted, and the Air Staff, and the world, and they bought it, and they're designing it."

Mitchell smiled. He said:

"The missile itself has been one of the least of the problems."

The Major Zorn and Lieutenant Colonel Mitchell were replaced at the table by a tall, blond, burly lieutenant who described the Strategic Air Command role in the world in tones measured to the millimeter. He used only one unusual term: at one point he described the United States arsenal of gravity-fall nuclear weapons. When a photo of these devices flashed on the screen it became clear he was talking about bombs. The briefer was interrupted only once, by the disconcerting failure of his projector light during a film clip of the F-111 aircraft, which is not known as one of the Air Forces most reliable planes.

He, in turn, was followed by a slender, dark-haired young briefer who delivered the Command Intelligence Briefing on the Soviet Threat.

"During the last decade," the briefer said, "the Soviet Union has been involved in a relentless and unabated buildup of its strategic forces."

At last Michael was taken to the SAC command center. Under escort he descended down concrete halls and stairways. "Please stand behind that line, while I check identification," said a succession of guards who wore white braid on their left shoulders. "Thank you."

The command post seemed familiar, although he had

never been there before. He had a collection of articles by writers who had been either enthralled or appalled with the place. It was an enormous room. Six huge screens with colored lettering and diagrams filled a two-story wall. Clocks displayed the time in places like Omaha, Moscow, Guam, and Zulu, which is Greenwich mean time. The escorting officer, another friendly captain, walked him through the routine. His name appeared on one of the screens; he heard a conversation with a SAC airborne command post "over the central United States"; and he listened to brief contacts with SAC bases in Point Barrow, Alaska, and Guam — "The weather here in Guam is a balmy 8o degrees. It is four-twenty A.M."

As they left the command post up the concrete halls, Michael's escort remarked:

"It is doubtful we would be around here more than thirty minutes after everything kicked off. We can get the message out that we need to get out in the time we have to survive."

II

Michael flew east. The great land stretched forever ahead and behind the aircraft, endless fields ripe for harvest. From above, the sheen of wind and sunlight on a hundred lakes made them look like blobs of mercury.

He stopped at Dayton, Ohio, and went to Wright Patterson Air Force Base, where the Air Force stores its history.

The museum was a huge Quonset hut of a building, surrounded by aircraft and missiles. On the lawn near the array of missiles two Airmen were training German shepherds. The morning sky was full of young thunderclouds, massing toward evening. Michael walked down the row of missiles.

Convair "Atlas," HGM16F.

Douglas "Thor," PGM-17.

Martin "Titan," HGM-25A.

Chrysler "Jupiter," PGM-19.

Boeing "Minuteman I," LGM-30A.

Boeing "Minuteman III," LGM-30G. Cost $1,818,000. Maximum speed, 15,000 mph. Range 8,000 miles.

Michael stood at the end and looked back up the row. The weapons were gray and white. They were beginning to look weathered. Threads of stain ran down their sides. Michael had his notebook open. He looked at them, and waited. The missiles stood quietly on the grass, passive columns of shaped metal. Across the expanse of lawn came the sound of the commands of the dog trainers. He could see the dogs crouch in response. He looked back at the missiles. Finally he closed his book without writing in it and went inside.

The building was jammed with machinery, photographs, and explanatory signs. He walked through the rooms at random. Here was a photograph of the first shot fired from an aircraft: August 20, 1910. Here was a replica of the Kettering "Bug," an unmanned biplane considered the first missile. Here was a nine-and-a-half-minute movie of World War One aviation. He spent time standing beside the first modern bomber, the Martin B-10, first built in 1933, "The Air Power Wonder of its Day." He glanced in passing at a display of spark plugs, 1936 to present. He stood for a long time in front of the photographs of the men captured by the Japanese after the first American bombing raid on Tokyo in 1942. He looked at a photograph of a burning United States A-20 aircraft starting its final dive. Next to it was a picture of a B-17 taken at the moment one of its engines was blown up by flak. Just across a hallway from it was a small, heavily retouched photograph of a German V-2 rocket standing on its pad.

The museum was almost empty. He wandered alone among the mementos and the pieces of hardware. There was a sound of fans and faint, unintelligible music. Here were the aircraft, enormous winged creatures standing on the immaculate floor. The Lockheed "Lightning," P38L; the Martin "Marauder," B26G; the Northrup "Black Widow," P61C. Near the aircraft a small back-projection booth was repeatedly running a short movie on the kamikaze pilots of World War II. Every time Michael glanced in that direction he glimpsed aircraft blasting into ships and exploding.

Here was the very B-29 that flew the Fat Man Bomb to Nagasaki and let it go. Michael stood at the side of the aircraft, again with his book open, but wrote nothing. He left the B-29, and went to look at the Inertial Measurement Unit for the Titan II, a nondescript chunk of metal that was slightly larger than a beer keg but did not resemble one at all because it was plated with gold.

And there was the bomb, near the circular kamikaze movie. Replicas of the bomb. Fat Man, Nagasaki. Little Boy, Hiroshima. Michael stood there, too, looking at the curious, finned shapes. Kamikaze aircraft crashed again and again on the nearby screen. Michael wrote nothing.

At last, he walked on. He hardly glanced at the enormous B-36 bomber, whose aluminum skin was so lumpy the aircraft looked as if the whole thing had been overinflated. He showed little interest in the yellow Bomb Lift Truck underneath, nor in the adjacent replicas of the Mark 41 and Mark 53 thermonuclear bombs. He paused, and then stopped, at a display of World War II artifacts.

Here was a small, off-round object that looked like a large rugby ball. It was torn. Its several layers of lining and a twist of broken thread protruded up through the ripped leather. The sign beneath said:

"Flak helmet worn by a T/sgt H. Farwell over France on June 15, 1944 when he was hit in the head by a piece of flak from a German anti-aircraft shell. (No further information is known of this incident but in all probability Sgt Farwell did not survive such a wound.)"

Michael stood beside the glass case for a long time, writing in his notebook.

At the door was a small stand bearing George Santayana's observation: "Those who cannot remember the past are condemned to repeat it." In the lobby a man in a colorful suit stood by a sign that advertised "Mike Stone's Amazing Dip-Er-Do Stunt Plane." Over and over the man threw a tiny model. Over and over the model looped back to his hand.

III

The most popular display at the Air Force Association Convention at the Sheraton Hotel in Washington was a model of an aircraft made by the Lear Siegler company. Young Air Force lieutenants and captains lined up to fly the thing around the ball, using a new kind of stick that responded just to the pressure of the hand. The little jet whirled, climbed, dived and twisted hour after hour.

Michael wandered through the display booths in the basement of the hotel, carrying his notebook. All around him mingled recorded voices filled the air with garbled enthusiasm. Back-projection booths caught the eye in all directions: aircraft in flight, aircraft landing, aircraft standing still, pilots climbing out of aircraft, smiling. He stood for a while beside a display by General Dynamics that consisted of a wall covered with about fifty 8 × 10 pictures of cruise missiles all photographed in flight against Southern California backgrounds. The land was familiar to Michael — deserts, grasslands, woodland; it was like seeing the sky of home filled with bombs. Nearby, a white robot sponsored by ITT Gilfillan rolled around the display floor flirting with young women while a gray-haired man twenty yards away smirked into a microphone hidden in his palm.

This display of hardware seemed to encourage excess. Most of the officers looked polished, trim and aggressive; the women wantonly gorgeous; the journalists enthusiastically cynical. There was a representative of the *Village Voice* sketching on a pad while a bald industry representative explained to him patiently the operation of the MX shelter, as modeled by Boeing. His drawings would not be complimentary.

Michael spent some time standing in the background at the Boeing display, watching the little model of the MX transporter back up to a horizontal silo, run its model missile back into the hole, retract a mass simulator shell, drive away, back up, insert the mass simulator shell, withdraw the missile, and drive away again, to the accompaniment of the buzzing of several electric motors. The transporter repeated the exercise tirelessly. The bald Boeing man was pompous, and endlessly patient.

Michael drifted over to the Rockwell International Autonetics exhibit. Inside a screening room the twelve slide projectors and the strobe lights clicked and flashed. Outside, a man Michael had met in Southern California told him: "They took out the picture of the skulls." Michael went in. The screen flashed with photographs of fire, stone faces, and smoke. "John Milton," the voice said somberly, "told us of that first war . . ."

Michael stopped at the Avco display, where a huge, full-size mockup of the MX fourth stage stood, dwarfing its admirers. It was cut away to reveal its forest of conical black RVs. One of the RVs was in turn cut away to show the interior of the new warhead Avco was promoting, the Advanced Ballistic Reentry Vehicle, designed to replace the General Electric MK 12A that was planned for the MX. This ABRV could trace threads of ancestry back to Marvin Atkins, Seymour Zeiberg, Big John Toomay, and the rest of the boys at ABRES, and it carried the brand-new Livermore bomb. Michael looked without passion at the cutaway.

". . . So we heard that some guys from the Soviet Embassy were coming over," one of the Avco men was saying, and Michael eased himself into the conversation. "They were going to take a look around. So we had this sign made up."

The man dug around in a stack of material and found a small black-and-white plastic label, modestly official. He stuck the label on the fourth stage mockup. The sign read: "One-Quarter Scale."

At the Northrop display there was a model of the Northrop-Draper Labs guidance system beer keg. Michael expressed a mild interest in it to a man who was standing near it looking left out. This central piece of hardware, the one device which, like a lens in the sunshine, could focus all the power of the rest of the weapon into the full exertion of its force, was ignored by the convention.

The man jumped at Michael, eyes alight.

"Here's the sphere assembly. The heat exchanger. The power shells. The equatorial ring. The liquid used has a neutral buoyancy; it's a hydrocarbon that does not conduct electricity; it's clear like water, but it's more dense than water; it's more viscous than water; it has the following functions: it minimizes thermal gradients; the atmospheric field stays the same; and there is no amplification of vibrations."

The man stood close to Michael, as if imparting a secret. He said:

"If you think about it: the shock, the temperature; a human embryo is protected in just the same way!"

Michael nodded and nodded, staring at that familiar ball.

The day, and the three-day convention, came to a slow end. Late in the afternoon Michael wandered back through the displays. The Lear Siegler magic control stick was now so completely desensitized to the pressure of human hands that it would not budge the little model of the jet even if you yanked on it. The Boeing model of the MX system had become stuck with the missile halfway into the shelter. Someone suggested dousing it with a bucket of cold water. And in the Martin Marietta booth, where several speakers had spent the three days talking continuously about the MX multiple protective shelter system, there was now quiet. Michael put his head into the little portable room.

Three men sat there talking among themselves: a Boeing executive (who had a full head of gray hair), a lanky man from Martin Marietta, and Seymour Zeiberg, bulky, crewcut, ebullient.

"Hi, Michael!" Zeiberg said.

"How are you?" Michael said.

The three men looked secretly amused. Possibly they had been discussing the latest rumor that Zeiberg, the old insider, had himself put into circulation at a cocktail party a few days before: That the newest MX system would consist of one hundred missiles and ninety-nine shelters.

"You see," Zeiberg said to Michael with a grin, gesturing at his companions. "I come down here to see all this stuff and I can't get past the first booth before I run into my cronies."

IV

Major General Jasper Welch watched Michael with eyes that looked sleepy but were not. They sat at a long table in Welch's office. On the wall behind Welch was a painting of biplanes, smoke, and flame, labeled "The Bombardment of Ostend by British Machines."

"These are positive feedback groups," Welch said. His voice was a lazy, confident rumble. "You see it in fighter squadrons, even. You take all the general officers and see what fighter squadrons they were in and you find that it is very nonrandom. You have to be, one time, in a red-hot group."

Welch seemed to be a thick man, without looking fat. He sat still at the head of the table, that mask of slight drowsiness pulled over his face. Big John Toomay had spent a lot of time talking about Jasper Welch: "He's not a very articulate guy. He lets his mind wander. He says things that, if you try to fathom them, you can't figure them out. But when I entered the Industrial College of the Armed Forces I took the standard test that compares you to all the other people who are entering graduate business schools all over the country. I got a score in the ninety-third or -fourth of all the guys applying to business school. A guy I was with got ninety-seven percent. I was talking to one of the instructors there, and I said I thought that was pretty terrific. The guy said, that isn't terrific: let me tell you about the guy who was here last year. He MAXED the test. MAXED it! His name was Jack Welch.

"That convinced me. Even now when I talk to Jack and he says these things I don't understand, I keep remembering that."

Jasper Welch got his Ph.D. in physics from the Livermore Laboratory in 1957. He had been Bill Crabtree's boss when Crabtree was a lieutenant at the Air Force's West Coast Study Facility. He had studied the future of ICBMs long before Strat X. He was part of the famous "B Team," a group like a Tiger Team that had changed the way the United States looked at the Soviet threat in 1976 and 1977. He had worked in the White House on the National Security Council during the Carter Administration. He had helped John Hepfer get Bill Crabtree out to Norton Air Force Base. When John Toomay worked in the Pentagon for Jack Walsh, who had the champagne parties, Jasper Welch sat at the table during policy discussions and made remarks that Toomay was later instructed to work into speeches. Welch had worked with Seymour Zeiberg on MX; once the two of them canceled a prized trip to the Paris Air Show to fly instead to Norton for a long weekend's work. Now he was Air Force Deputy Chief of Staff for Research, Development and Acquisition. But he traced his own career back to incandescent days when he and Marv Atkins and a bunch of other future stars had worked at the Kirtland Air Force Base Special Weapons Center on nuclear weapons effects. That, Welch said, was his red-hot group.

"The feedback mechanism in these groups is really pretty remarkable," Welch continued. "You get around good people, you get good habits, you get good insight, and that leads to other things, and that's why, for example, of the Kirtland Mafia, out of the thirty-six of us who went to work there on Christmas Eve, nineteen fifty-seven, there are now twenty-one general officers. And that doesn't count the guys who have general officer equivalent positions in the civilian part of government, or the two who are vice presidents in industry."

Welch chuckled. His sleepy eyes became amused at the memory. Outside his office it sounded as if several men were throwing garbage cans out a window. Welch was unperturbed. Michael asked him where he got to know Albert Latter.

"He was very influential when I was at Kirtland because he was on the Nuclear panel of the Air Force Science Advisory Board, which was the guru group the laboratory dealt with. We got to know them very well. When I left Kirtland in nineteen sixty-two, I went and worked for Albert at Rand. So the interconnects are very strong."

The noise outside had stopped. Welch's voice was quiet in the large room. His eyes were on Michael, but it seemed as if a part of his thought was elsewhere. Michael began to stuff his notes together, preparing to leave. Among them was a transcript of John Toomay's praise of this unusual man:

"He's not a very good military guy. He doesn't dress very well; he doesn't keep his hair cut very well; he doesn't always shine his shoes; he lets his socks hang down; when he visits the chief he puts his feet up on the desk and all that; he's really informal. But his mind is voyaging out there somewhere where none of the rest of us can ever go."

21

I

THE building that housed the corporate headquarters of the Martin Marietta Company in Bethesda looked like a machine-gun emplacement. It seemed to be made of a little tinted glass and a lot of gravel. It was a massive block of a building with a surface of glossy stones. It rested in tiers on a huge expanse of lawn in which saplings grew. From the outside its windows were black. Across the empty boulevard was a building occupied by the International Business Machines Data Processing Division.

Inside, Michael found himself at the bottom of a shaft of muted light that hung between a ladder of balconies on one side and a wall of tinted glass on the other. Beyond the glass the lawn and trees had all now gone deeply too green, as if they had been spray-painted. The inside walls of the building were also faced with gravel. The impression was of hardness, but not solidity. Michael introduced himself to a receptionist: "I have an appointment," he said, "with Dr. Seymour Zeiberg."

After a few moments, which Michael spent glancing at

a copy of *Aviation Week and Space Technology* that rested on a table, he was led upstairs to an office off a broad corridor. In accord with an odd trend, secretaries worked at desks placed just outside their bosses' doors, in the hallway, giving the impression that they had been recently thrown out.

"How are you?" Zeiberg said, with his old grin, both shy and challenging. His crewcut stood boldly on his head, still dark gray.

The office was small. On one side there was a modest brown couch. On the other side, near a tinted window, was a desk of modest size. There was a modest stack of papers on the desk. There was a modest round table at one end of the room. The secretary, Bea, sat modestly out in the hall. Zeiberg, large, broad-shouldered, thick-armed, wearing a short-sleeved blue shirt and that aggressive crewcut, seemed out of place. The only thing in character was a brass gong that hung from a wooden stand on his desk. Michael was familiar with the gong. It had a place in history beside the grenade and the bull-shit repellent: Sometimes, in briefings, Zeiberg had been known to beat on the gong until the unfortunate briefers faded away in shame.

Michael grinned back at Zeiberg, and shook his hand. He said: "I thought you had wanted to go back to Southern California."

"Yeah, I did," Zeiberg said, then smiled again. "But I had an offer I couldn't refuse."

Zeiberg was now the vice president for engineering for Martin Marietta's Aerospace Group. His power had significantly diminished. His income had significantly grown.

"I'm trying to get acclimated to the company," he said. "I'll be spending time in Denver and Orlando. I'm getting some new furniture. I have a big desk on order."

"Somebody said Reagan is going to announce a decision on MX next week," Michael said.

"That's what we hear. Although it is not an issue he should highlight. No matter what he does he's going to get a lot of flak, so he ought to submerge it in other things. Flood the public with these decisions."

Michael looked at Zeiberg curiously. Zeiberg sat leaning back in the chair, his hands comfortably behind his head.

"I was told that the hardest thing you have had to do is persuade people you had — looked at all the options."

Zeiberg nodded. He said:

"Perry and Brown were both so hopeful on that airmobile thing."

Zeiberg rumbled on. Michael sat back and watched the reels of his tape recorder turn. Deep in the Martin Marietta bunker he listened to the stories of the second full winter of the Carter administration.

The winter of 1978. The Pentagon. D-Sarc II for the MX. Hepfer, Crabtree, Toomay, Perry, Zeiberg. Back down to 3E801. Perry: "That was a best-seller, that D-Sarc. We had the unpopular task of limiting attendance." Big table. Air Force on one side, Defense Department people on the other side, briefers up front. Crabtree and Hepfer have already briefed almost everyone in the Pentagon. By now the story is old. Perry has been spending days at Norton. He runs the D-Sarc, but for him the story is old, too. The story is Vertical Silo, Multiple Protective Shelter. Move the missile from hole to hole with a Transporter-Emplacer, infrequently. Lowest cost, quickest IOC, easiest to hide. Perry: Thank you, gentlemen. A fine presentation. Your work has been absolutely excellent. Thank you.

Crabtree: "Before we got to the D-Sarc — it was already — fairly clear that the President was not going to accept — the vertical shelter recommendation."

The problem apparently is the President's scientific advisory committee, led by his adviser Dr. Frank Press of MIT. His group doesn't think a vertical silo shell game is a very good idea. His group thinks that the Soviets, A, won't be fooled enough, and, B, will just be fooled enough to be unable to count the missiles from space, the way it says in SALT treaty language. They think putting missiles in airplanes is still a pretty good idea. And the President is listening to Press.

So what do we do with this difficulty?

In the Martin Marietta building, Zeiberg folded his burly arms. He continued:

"We were brainstorming with Brown, asking what the hell do we do with these guys? They had the President's ear. He wanted to hear what he was hearing from them, and he didn't want to hear what he was hearing from Brown, obviously. I suggested that, they had come out so strong for airmobile, we should study that. So I devised a twist on airmobile. Conceptually it was really nifty. It had all the features of the multiple aim point system. I devised a scheme where the airplane which had the right short take-off and landing capability could use the general aviation airports, and each one of those austere airports could act as the equivalent of a shelter. You could access wherever you wanted to, then drive a small vehicle with the missile off the airplane."

A new basing mode! Spring, 1979. Bill Crabtree: "They told us to press on — with the vertical shelter — but in the meantime to do some study work — on a new form of — airmobile. A form that we came to call — air mobile MAP."

Norton Air Force Base. Rain. Wind. Warnings to trucks, trailers, and campers for gales on the highway out of San Bernardino north. Bright air. Instructions go down to Crabtree, and the word goes out into the labyrinth of the Ballistic Missile Office. Money is spent, and jobs are created, and a man named Harold Kinsock, whom people call Bud, develops a survey of all small airports, large airports, stretches of straight highway, dry lakes, salt flats — anything on which an airplane with a missile can land, open its hatch, and roll out its deadly cargo: BLAM!

Hepfer: "Here they were talking about landing on runways that were twenty-five hundred feet long! With a four-hundred-thousand-pound airplane. I thought we'd fix that in a hurry; the SAC pilots would tell us we were crazy. They said 'Oh, we think we can do that.' "

Little John Hepfer and Big John Toomay, over at Systems Command, keep an eye on the airmobile study, and insist on careful work. But when it comes down to it, they tell the world how they feel. Congress, particularly. Congress gets curious, and military and appropriations committees start having hearings. Hepfer is an occasional witness. One day Senator Barry Goldwater spots him.

Goldwater: "General, would you want to be IN one of those airplanes that landed on that 2500 foot strip?"

Hepfer (an inevitable grin): "No, Sir, I really wouldn't."

On another occasion the presence of Big John Toomay has been required. He sits all day, waiting to be called. At last the question is put to him. "General Toomay, do you have any experience in the strategic world?"

Toomay: "Yes, sir, about twenty years."

"Well, what do you think of airmobile?"

Toomay considers a moment then bursts out:

"I think it's dumb."

The next day Toomay is called in by the Chief of the Air Force for the inevitable and impossible request. Say, John, the next time you go over there to the Hill, don't just blurt out something like that. Try to be more diplomatic.

Seymour Zeiberg sat comfortably in the chair. He folded his hands on his belly. He grinned. He went on.

"Perry and Brown both were so hopeful on airmobile. Brown thought that would solve all the political problems. You could almost see him sitting with his fingers crossed. Perry, too. But it was expensive, and the operational guys never felt comfortable with it. Then we had to give up on the vehicle getting off the airplane, because for a reasonable size airplane the missile had to be small, and there was a political desire for a big missile. Then the arms control guys kept voting no on the vehicle because of SALT verifiability. Then we found out that those airfields weren't all that easy to use in the current state. You wouldn't want to land a seven-hundred-thousand-pound airplane with ten nuclear warheads on some of those strips. All of a sudden we had five to ten billion dollars of airfield improvements in addition. All in all it came out to be about twice as expensive as any other option."

OK, here we go, back downstairs. Back to 1E801. D-Sarc IIB. March 31, 1979. An all-day Saturday D-Sarc. The airmobile idea is gone. What's to take its place? Bill Perry: slender, sad, almost spectral, shaking his head. Alas.

Gentlemen, the Soviets are building — Gentlemen, the Soviet Union has now fired BOTH its SS 18 and

SS 19 Intercontinental Ballistic Missiles using its new guidance apparatus. Gentlemen —

Crabtree: "Subsequent to the D-Sarc IIB — the Department of Defense prepared itself to take a decision package forward — to be reviewed by the various decision-makers. My understanding is that variances of the vertical shelter — of airmobile — of strengthening other legs of the Triad — and do nothing — were all options that were taken forward. I — of course — was not very privy to that decision-making process — but it seems clear that none of those options — were acceptable."

Zeiberg, however, was privy. Now he sat back and grinned again. There was a little wolfishness in his grin, behind the layers of aggressiveness and uncertainty, a predator's eye. A friendly predator; a kind of family-man predator.

"It was pretty unruly the last few months," he said. "Things were, in a sense, unstable, because there was strong opposition to doing anything. The President was saying it was necessary to do something; Cyrus Vance didn't want anything he didn't believe was needed; Stansfield Turner didn't think it was needed or desirable. Arms control based opposition was coming in, and anything we proposed was not passing the screens that were set up for verification. We went through a furious few months of generally every two or three weeks changing the design."

Zeiberg sat at the table, composed. He was entirely unlike his early predecessor in this position, the notorious Lloyd Wilson. Toomay: "Wilson was seen as enormously powerful because he TOLD people what to do. Zeiberg wasn't like that; he was willing to make compromises, so he was not perceived as exercising his power." Too, Zeiberg had not met a romantic end. Reminded of Wilson, Zeiberg grinned. "Lloyd was quite a character," he said, then went on, reiterating the most memorable characteristic. "You know he finished on the receiving end of five short thirty-eights."

But Zeiberg's tenure in that position had not been placid.

The Pentagon, in the spring of 1979: Gentlemen, make haste! Alas, beware. Gentlemen, make haste! One Saturday night during the process Zeiberg, Perry, Atkins, an Air Force colonel, and Zeiberg's secretary, Sandra Van Namee, gather around a copying machine assembling a report while the voice of Harold Brown comes filtering up to them through channels.

Zeiberg: "The report had to get to the President Monday and Brown wanted to review it Sunday. We're cleaning up this massive report, and Brown's getting antsy, and he gets very snappy when he gets irritated."

Harold Brown, this analytical, brilliant man, will devour the report in a few minutes, once it gets to him. But it isn't here. It isn't here, and the decision is waiting. The decision is upon us! "Where the hell is the report?" In the humming room upstairs the undersecretary, the deputy undersecretary, the assistant deputy undersecretary, the aide and the just plain secretary carry sheets of paper to the machine, which sucks them up and spits them out. Airmobile. In Line Openable Trench. Option Three. Sandy, will you? Option Four. Page Sixteen of Forty Two. The rationale for the use of the hardened vertical silo hinges on estimates of CEPs of . . . Part 6. "Where the hell is the report?" Page 28 of 200. The Mostly Common Missile and the Common Missile. Arguments in favor of SUM. The rationale for the strengthening of bomber and submarine forces as a replacement for the traditional Triad . . . Sandy, can you go get? Part 8. Part 2B. Horizontal Silo Hardness. Preservation of Location Uncertainty is based on several factors: . . . Verification. "Where the hell is the report?"

The copying machine breaks down. Lights in the wrong places. Crumpled paper. Marv, do you know anything about these things? What if we? It'd melt. Here, pull this thing out and try it again. Needs paper. No, needs ink? Here, push the button on the left. Tilt! Check the paper path. Look at this stuff on the roller. Oh — Thanks, Sandy. Is there another one upstairs? Marv'll have it fixed here in a minute. Yes. Ouch. There. OK, close it up. "Where the hell is the report?" Can't we give him something else to do?

In his office Seymour Zeiberg has another device. He does not use this one to massacre the pride of briefers, but it might come in handy for a kind of opposite purpose. It is a model of a new idea which he asked the Boeing Corporation to whip up. He calls it a Cruise Ballistic Missile. It pops out of the silo, develops wings, turns into an airplane, flies off, then shoots a missile. Great concept. Atkins, laughing: "The cruise ballistic missile. ICBM. The famous model. As it's flying around if you want to shoot it the wings come off, otherwise it comes in for a safe landing. In principle."

But there's the model, gathering dust near the grenade in Zeiberg's office. Perry, somber, with a stack of papers still to be collated, says to Zeiberg, "Why don't you take that model down and show it to Harold?"

The lair of the Pentagon King. Brown feels that he can shout at Perry but has to be polite to more humble subordinates. Noise at the door. Where the hell is — Hello, Sy. Have you seen this thing yet? Hmm. Cruise Ballistic Missile? See, this is how you put it together. Then this is how it works. Model leaps out of silo, sprouts wings, flies around the room, shoots missile. Hmm. The idea is. It works like this. Leap, sprout, fly, BLAM! Hmm. Cruise Ballistic Missile. Boeing. (Upstairs, collation continues.) Hmm. What if the? Yes, we could certainly do that. Leap, sprout, no BLAM, land — million dollars each. Yes, hmm. Entirely feasible. Based on an idea we. Delayed reaction. Slow. Fast. Leap, sprout, fly — Sy. Yes? Would you go up and ask Bill: Where the hell is the report?

A long time later, when the MX mafia is splitting up and Bill Perry is leaving the Pentagon (alas, farewell) for his job with Hambrecht and Quist, Perry holds a little ceremony in his office and gives Defense Department decorations to Zeiberg and Atkins among others. Each man makes a few remarks appropriate to the occasion, some somber. Atkins, when called, gets up and says:

"I do remember one time that Bill Perry put me under an almost unendurable pressure."

There is some curiosity in the room, because Perry's gentleness is a legend. This somber, slender man with the sad eyes simply does not lash out, does not seek to bruise. He looks at Atkins with an embarrassed curiosity.

"I remember," Atkins says, "one Saturday afternoon when we were working at the Xerox machine. Bill Perry came up behind me and said: 'Marv, can't you make it go any faster?' "

The decision is upon us! In remembrance, the strange, half belligerent, half apologetic smile again appeared on Zeiberg's face. "It was a taxing day," he said.

Michael looked at him. There was a moment's silence. Michael rushed to fill it, out of context:

"Where do you put yourself on the political military spectrum . . . Hawk — Dove." Zeiberg smiled again.

"Like something Brzezinski said once, he's a dawk."
He laughed. It seemed to be a gathering laugh, a calling
in of the forces of the man from their various amusements
into an array of power. His substantial body became still
in the chair. He put one solid hand on the table. But his
voice assumed no new weight; if anything, it was more
elaborately casual.

"I guess in a sense I'm that. I'm hawkish in the sense
that I believe that the Russians are bad guys aimed at
taking whatever they can, and consequently we have to
have a strong military because they respect that. If you
read what they write and understand the structure of their
empire, it's dominated by political control made possible
by a strong military organization, and that's what they
appreciate and that's what they respect. I don't trust their
motives and I don't think we'll ever convince them on
intellectual arguments or moral arguments or whatever
of any steps towards the direction of arms reduction. By
the same token I think that arms control and reduction
are the only answer. In a sense I am a dove in that I
believe arms control will be essential, but the problem is
I don't see how you can make it happen until it becomes
financially painful. For them, too."

"Have you developed a sense of the Soviets from any
particular readings — any particular sources?"

"Yeah. Reading miscellanea. My family is Russian,
so —" He grinned. Not apologetic, this time. Suddenly
Zeiberg's face, and his stocky body, looked like an ad-
vertisement Michael had seen in *Aviation Week and Space
Technology* for pop-up training targets. The targets were
clearly ethnic: they were of stout men with broad Slavic
faces, each bearing an AK-47 rifle with its distinctive
curved magazine. Ivan. He leaps from the brush again
and again, and again and again you shoot him down.

Ivan! BLAM! Ivan! BLAM! Petrov! BLAM! Ilya! BLAM! Sy!

"So," Zeiberg continued, "it's not really instinctive, but having met a lot of people who —" he stumbled. "— not anymore — even relatives, a different attitude you see in the people from the Soviet Union versus people from England, for example. You also have to — the more you get to know about their military capability and studying their exercises and operations, watching the developments in their acquisition programs, we develop a model of what they're up to. Then they say a lot of things which are ignored by most people.

"I think the fundamental problem we've had in the country — still have — is a tendency to think the Soviets are like us. Poor understanding of the Soviet government, the Soviet policies, philosophies, have misled us to the extent of some people advocating unilateral disarmament. That is, I think, a cop-out. Imagining that the enemy has the same attitude as we do. That's not the case. The Soviets are different people, different history, different motivation, and unless you want to try to make an assessment of that, you're just fooling yourself."

"Where do you get this feeling?"

"Most of the material I've seen comes through government translating. Government-sponsored translating operations. The Air Force has a very extensive operation and their foreign technology division dedicates considerable effort in that."

"You don't remember any particular . . . sources — ?" Zeiberg's answer was quick.

"There's no one thing that strikes you. There is so much that it comes through and it all aggregates in creating an impression."

Michael looked hard at Zeiberg's broad, steady face.

He wrote in his notebook: "Trust me!" He looked up into Zeiberg's eyes, but his own big eyes gave nothing away.

II

Early June, 1979. The Pentagon. Seymour Zeiberg's office, sometime in the middle of the day. There's a meeting of the National Security Council tonight! Carter meets with Brezhnev next week. Zeiberg: "We needed to do something strong."

So, we need an update on the choices, an accurate update! Something you can see at a glance. Something graphic! Some graphics! This afternoon!

Sandy — do you have any poster paper? Poster paper? What about some felt-tipped pens? You know, the kind with the thick end. Posters, like in high school, for Christ's sake! "Go, team go," you know, "Beat the Bears." Who would have those things, in the Pentagon? Nobody makes posters in the Pentagon.

Any hope, Sandy? Got the paper, working on the pens. Working on it. Great! Meeting's coming up.

Zeiberg: "If the Marx Brothers had written the script ... Here we are, a critical meeting with the President, all these years of knowledge and experience on the subject. A mountain of material available, and we're hand-drawing the charts. It contributed to the general insanity of the process we were going through."

They gather around the long table in Zeiberg's office. Atkins, Zeiberg, Perry, Welch, an MX colonel, and the secretaries. They stand around the long table, preparing charts. Perry is the architect. The others operate the markers. Sandy Van Namee tapes the paper together as they finish each poster. They write in bold black letters across the white. They sketch missiles that look like Fourth of July rockets. Options for the President.

"Track Mobile, with open connectors."

"Track Mobile, with aboveground connectors."

"Track Mobile with Openable Trench."

"MX Missile, Mostly Common Missile, Common Missile."

"Vertical Shelters."

"Horizontal Shelters."

Accuracy. CEPs. Ballistic Missile Defense. Verifiability. Cost. All the years of study and argument, sketched out on eight sheets of white paper. Strat X. Jim Drake waterskiing. MIRVs. ABRES. Seymour Zeiberg and Big John Toomay flying to Washington on the red-eye to struggle for funds. Lloyd Wilson. Incredible. WS120A. The rail mobile Minuteman, a distant memory. The Minuteman dropped from the C5, a forgotten exercise in technological diplomacy. In many ways this is just the same thing, on a grander scale. Thiokol. Martin. Rockwell. Pave Pepper. The AIRS guidance system, spinning its way to perfection in clean rooms in Massachusetts and California. Albert Latter, the restless one, the inventor. The trench, abandoned in the desert. Bill Crabtree, solemn and intense, riding the red-eye to Washington to brief the entire Pentagon. The MX Transporter, the largest rubber-tired vehicle in the world, grinding on its many tires across the Nevada sand. All summarized in the authoritative block writing of the men who saw all these things come and go, and who persevered. Two years later Zeiberg, leaving the Pentagon, as Perry had done just ahead of him, will ceremoniously present the famous handwritten charts to Marv Atkins, in a moment of laughter and nostalgia, but today as the small group finishes its sketches and their summaries, Bill Perry collects them up, smiles gently, sadly, and takes them away to show to a President who, just two years ago, condemned the nuclear weapons arms race and pledged to move toward the elimination of all nuclear weapons from the earth. The decision is here!

22

I

MICHAEL walked on the road between the monuments. They rose at regular intervals: eagles, warriors, tablets, arches, goddesses, men on horseback, and obelisks, obelisks, obelisks. He carried a green and yellow map in his pocket, to tell him about the scenes as he passed them. "Along and in front of this ridge, Lee marshalled his troops . . ."

The fields were unmarred. The woods that had provided shelter for soldiers were groomed. There was no deadfall. Between the monuments, along the roads, and in the woods, were cannon, pointing silently at each other. Michael walked between them. On the horizon to his left rose a huge, slender tower, a private observation tower erected for tourists. Michael had been reading a book published by the United States Department of Defense called *The Effects of Nuclear Weapons.* It would not take more than a pound or two of overpressure to turn the tower into a shambles of wire and steel. The monuments would survive. So would the barrels of the cannon, pointing steadily across at each other where they fell as their wooden wheels burned away.

Monuments everywhere. "In memory of the men of the 12th Regiment of New Jersey Infantry Volunteers. This monument is dedicated by their surviving comrades as an example to future generations." "The First Regiment, Second Brigade, Third Division, Second Army Corps." An obelisk "erected by Congress in honor of the Cavalry, Artillery, Infantry and Engineers of the Army of the Potomac." An obelisk that looked like a conical RV, "the 14th Vermont Volunteers."

"Fought with honor." "Fought with honor." Each monument stood a few hundred feet from the next; in their regular spacing they looked like tombstones. It seemed that they marked the graves where these entire regiments had been interred, where all had been killed instantly, leaving a battlefield one moment alive and screaming and the next as silent as it was today. Here lies the First Pennsylvania Cavalry Brigade, Second Division Cavalry Corps, Army of the Potomac, who fought with honor on the fields of Gettysburg, who died together on the third of July, when the great fire came.

Michael walked up the steps of the grand stone monument to the soldiers of Pennsylvania. It was a massive structure, an arching shelter for plaques full of names. Here and there among the names one or two had been polished recently and stood out from the oxidized plaques, outlined by the dried Brasso, gleaming with a love handed down through generations. Samuel Ewing. Nathan H. Jones. Private C. C. Kirchner. In a few places other names had been erased, the relief ground away entirely. What dishonor swept them from the rolls of the brave and the dead?

At the edge of the road a deer had left tracks that cut deep in the sandy yellow dirt. Here was the Copse of Trees, where, on the afternoon of July third, a piece of a Confederate charge of twelve thousand men had reached almost to the high point of a ridge before being thrown back. With nostalgia, the brochure and all other interpretive material spoke of this place as the high water mark of the Confederacy. The place where that tragic alliance found its symbol and its greatest accomplishment in the futile shedding of its own blood. The copse was surrounded by a black metal fence.

Michael turned a corner and passed the Wheatfield, where repeated fights had raged, fights which, the brochure said, "left these fields bloodsoaked." Then the Peach Orchard, and more monuments. Fought with honor. Fought with honor. Michael walked up the stairs of an observation tower, the steps clanging beneath his feet. The view was all cars on highways, distant motel signs, and the other tower. The monuments had almost vanished. The cannon had disappeared. He jogged back down. Now he was on the Confederate side of the valley, where the army on the night of July second had looked across the lines, with only a few more hours of flow left to its tide, at Union fires. The monuments were now all headed C.S.A. "Alexander's Battalion." "Rhett's Battery." "The Brooks Artillery." "Losses heavy but not reported in detail." "Floridians of Perry's Brigade comprised the second, fifth and eighth Florida Infantry, fought here with great honor as members of Anderson's Division of Hill's Corps. 445 casualties of 700 present for duty.... By their noble example of bravery and endurance, they enable us to meet with confidence any sacrifice which confronts us as Americans."

Michael walked to the Virginia Memorial. On the brochure it was number eight. "General Lee watched the gallant charge of July 3 from here. And when it failed, he rode forward to the fields in front of you and rallied his men." The field was smooth. Cannon pointed down across it. The monument was a monolith of rock with a horse standing on top of it and Lee sitting on the horse. Down in front of it, with their backs to the stone, were the statues of seven men. A sign explained them. They were symbolic of the civilians who fought for the Confederacy: "a professional man, a mechanic, an artist, a boy, a businessman, a farmer, a youth." One was opening a cartridge to load, one held a rifle, the artist was shooting a revolver, one was swinging a clubbed rifle, and the youth was blowing a bugle, strain and fury in his face. Their heads were marked with a fluid green stain that flowed down into their eyes.

Michael stood in front of them for several minutes. A wind was picking up. He began walking around the monument to the north. He inadvertantly passed through the shadow of the top of the statue, the shadow cast by General Lee. He looked up. An approaching storm had cast its haze of moisture across the sky and had encircled the sun. He looked up and he saw the tall black figure of Robert E. Lee haloed in light.

He walked on. Cannons in the woods. Cannons peering out through the trees. One cannon, probably moved as a prank, pointed right at the center of the trunk of a substantial tree. It made it seem as if the cannon had stood there all these hundred years, as the tree grew, this one cannon watching the field where it had watched that day as Pickett's charge had been repulsed on the shreds of that force, four thousand of the twelve thousand who had left this slope only a few minutes before, had turned to escape the cannisters of death flung down from the doomsday cannons of the Federal forces on Cemetery Ridge. Monuments: "C.S.A. Army of Northern Virginia. Hill's Corps, Heath's Division, Pettigrew's Brigade. 11th,

26th, 47th, 52nd North Carolina Infantry. Present first day about 2,000. Total casualties 1,405." "To the eternal glory of the North Carolina soldiers, who on this battle-field displayed heroism unsurpassed, sacrificing all in support of their cause. Their valorous deeds will be enshrined in the hearts of men long after these transient memorials have crumbled into dust."

Michael walked into the woods. In the visitor center he had seen a painting of a girl with a full and childish face. Her name had been Jenny Wade. She had been twenty years old. On July second a stray bullet from the battle had come into her house, passed through two doors and hit her in the back of the head. In all this battle, through all the blood, the death, the hand-to-hand murder, the near breakdown of all of civilization in the heat of spilled life, what the brochure called the greatest battle ever fought on American soil, she had been the only civilian killed.

II

Michael walked through the shaded streets of Orlando, Florida. A brown van passed him at high speed. Ahead of it a tiny child, a girl of about two and a half, ran toward the road, wearing a bathing suit painted with tiger stripes. She came to the curb. The van approached. Michael opened his mouth to shout, to cry. No sound emerged. The van hurtled down the road; the child paused; the van passed her, its speed undiminished, its breeze ruffling her hair. Then she ran across the road.

Michael ran over to her. She stood on the grass beside the pavement. The street was otherwise empty. There were houses on all sides, but no one else was in sight. She was tiny. "Be careful," he said helplessly. "Be careful."

She looked up at him without comprehension.

Michael walked away.

At length Michael arrived at the door of General Bruce Holloway's house. It stood near the shore of a lake, under trees that shed narrow leaves in the warm Florida breeze.

The house was faced with brick; there were modest columns at the door. Welcomed in, Michael found himself stepping cautiously around the edge of a large blue rug that occupied a third of the living room. On the rug was embroidered a simplified version of the Strategic Air Command crest: A mailed fist grasping a handful of lightning.

"That was a gift," Holloway said. "It was made in Taiwan."

Holloway was very slender, a blade of a man. He offered Michael a seat on a couch, and sat down himself. Souvenirs of Holloway's career surrounded him. On the mantel behind him was a bronze horse, its mane and tail flying. On one wall, above the SAC rug, was the head of a bighorn sheep. On the floor next to the rug was a huge brass Turkish heater. It looked like a samovar. Behind Holloway's chair, where it seemed to be in reach if needed, was a sheathed saber. The saber looked like those in display cases at Gettysburg.

Bruce Holloway was the commander in chief of the Strategic Air Command from 1968 to 1972. The position is known as Cinc Sac. In 1971 the Strategic Air Command had taken the first official step in the development of the MX missile by formally expressing a need for the weapon. This document bore the official name Required Operating Capability; Michael called it an R.O.C.; Holloway called it a Roc. Bruce Holloway, therefore, was the man who had first formally asked for the MX missile. He had been the man who had visualized its use.

"What I was interested in," Michael asked, "is the

difference between what you asked for in nineteen seventy-one and what is now developing, as a weapon."

"I don't think there is very much difference," Holloway said. "The idea, in the Roc, the principal idea, in general terms, was something that would give you a higher order of reliability and greater acc'racy and be more versatile — and higher yield. That was about the package, and that's what MX does."

"Yeah," Michael said. "Now, at that time —" But Holloway broke in:

"But none of — no idea of this multiple protective shelter system," he said with feeling. "There have been some terribly wild schemes proposed. And the wildest one to me and of which I was the most vocally critical, was one that Al Latter came up with, where he's goin' to move missiles around *inside tactical warning!* Just preposterous!"

"You know Latter?"

"He's a —" Holloway made, then withdrew a vigorously critical remark. "I mean, those ideas were just, to me at least, out of the question!"

"There have been mobile forms of just about anything you can mention that have popped up since then," Holloway said. Michael smiled, agreeing. "One that had the greatest support was airmobile. I waxed rather enthusiastic about that at one point, but have since backed off. You would have security through mobility — but the horrible cost just turns your hair up on end."

"Yes," Michael said, and mentioned Perry and Zeiberg's excursion into airmobile options in the early spring of 1978.

"Right," Holloway said. "Another reason I don't like it is the loss of acc'racy. And acc'racy is a very sore subject with me. I've long been a strong advocate of gettin' all the acc'racy you possibly can in ballistic missiles. I'm pro anything that promises to improve acc'racy. It used to just *smear* me the way people, particularly in Congress, would fight against improving acc'racy because it was provocative — and destabilizing. It just makes you sick at the stomach. That's like saying if you're goin' hunting, after — whatever — a squirrel, a goat, or a lion — whatever — it's kind of unfair to go with an acc'rate gun! A weapon is a weapon, and why not make it as acc'rate as you can make it?"

"How do you feel about the debate over the accuracy of the MX providing a first strike capability for the United States and that whole question of —"

"See," Holloway said, leaning forward. "There's so much semantics mixed up in 'First Strike' that you have to be careful. Some people regard first strike as preemption in the face of *overwhelming* evidence that you're about to get hit. Some regard it as something as: Now's the time that we should go and let 'em have it. In that category I'm definitely against it. I've always been for our policy of never striking first."

"Yes. Yes." Michael nodded vigorously.

"But. You have to qualify that. If, again, the evidence is overwhelming that you're about to get hit the advantages of preempting under those conditions are very substantial. It might make the difference between surviving in some form you might recognize and not surviving in any form you'd recognize."

There was a moment's silence. Michael's pen hung over the page, frozen. Finally he said: "These things have emotional values now . . ."

"Oh, yes." Holloway said solemnly. "There's a lot of emotional aspects to — to particularly the targeting system that makes the most sense, too, and that's command and control.

"If we have a big advantage over them, to me it's in the flexibility of being American. Theirs is a rigid society. Nobody trusts anybody else. The command control structure is large and ramified and hardened and all that —" Holloway leaned forward. "— but just the same, *every* person in there is *scared* to death of exceeding his prescribed authority or taking over when somethin' goes wrong, and I just think that rigidity is a great weakness. And that's what I meant by targeting the command control system."

"What other targeting system makes sense in nuclear war? Ball-bearing industry doesn't make any sense, at least initially. None of that stuff, that World War II stuff, and that's the danger, if you carry that over, in an Iron Bomb war. Because it's going to be over!

"People say that's so ramified. They've got alternate command posts, hardened command posts — all the way down to battalion level. Just the way we do. They're protected by layers. I say: I know they do. You're not gonna get it all. You're only gonna get a little piece of it. I'm talkin' about gettin' a piece as high up towards the apex as you can. The apex itself and the layer immediately under it. And if you hit enough of it, and high enough, the whole thing is going to unravel, because of this rigidity."

"Just — knocking out the keystone," Michael said. Holloway rolled on.

"And it's so important! I couldn't believe more firmly in anything than that, with respect to nuclear war."

"And do you believe there will be an exchange . . . or do you think that we can avoid it?"

In the large room there was silence. Then Holloway said:

"I don't think there'll be an Armageddon war," he

said very quietly. "But I'll put it this way. There has never been any weapon yet invented or perfected that hasn't been used."

The telephone rang in General Holloway's home. Michael flinched. Holloway went to answer it. Michael, looking around the room at its striking contents — there was a very large bird mounted above a table in a sunlit extension — noticed a back issue of *The New Yorker* on the coffee table. When Holloway returned Michael asked him:

"Did you read the long story on nuclear war in there?"

It was a part of a book called *The Fate of the Earth* that was already becoming famous. It was a denunciation of the apparent drift of the world toward nuclear conflict. He had seen it quoted in every anti–nuclear weapons article, and heard bits of it at every anti–nuclear weapons rally.

"I didn't see it," Holloway said. "I hardly ever read that magazine. Look at the jokes sometimes."

"Oh."

Holloway pushed *The New Yorker* aside. Underneath, as if it needed concealment, was a magazine with a stark white cover on which was printed five dark blue stars and a red title: *Strategic Review*. He gave Michael a copy.

"You might want to subscribe to that," Holloway said. "I think that's the best defense periodical there is. I've been involved with that ever since it started, about ten years ago."

Michael glanced through the magazine. He had seen *Strategic Review* before. In an earlier issue it had shaken the defense community and sent waves out into the general press with an editorial by Arthur Metcalf, the magazine's military editor, arguing that the Soviet threat to Minuteman missiles was based on faulty logic: The numbers of warheads might make the threat seem real, but the reality of atomic war — what Holloway called the confusion of war — made it impossible for the Soviets to rationally consider the option. It had been the first attack on the window of vulnerability from the hawk side of the debate. The notion of the threat to Minuteman had been seriously damaged. Basing modes were less important; suddenly the MX missile was on its own. Might as well just stick it in fixed silos and call it an offensive weapon. From ABMs to multiple targeting, the same thing happened to MIRVs.

"Did you have anything to do with that editorial?" Michael asked.

"Yeah," Holloway said with a smile. "Arthur's an old friend. I collaborated a little bit with him on that, but very much encouraged him in the fundamental proposition. The question is, is Minuteman really goin' to be hostage? It all devolves back to this: I don't believe the acc'racy claims —"

"For the Soviets — ?"

"That's right. When you fire a massive number of 'em, which you've never done before, at targets across the North Pole, which you've never done before, you're just gonna have — it's just Murphy's Law, or somebody's law. Things are not gonna to be anywhere in the league of being as good as you claim if you've only done it on paper. And the biggest factor of all, is — whatever you want to call it — the confusion of war.

"It goes back to the shock effect." Holloway moved on. "I've always been in favor of at least doing more thinking and studying and planning with respect to limited nuclear options. I think the shock effect of just one weapon, say, put right in the middle of New York City, would be so *awesome* that there either will be — whoever it happens to will give in or the people that do it will notice the effect and feel that there is no way to go further without their own destruction — or somethin'."

There was a short silence. Holloway smiled. "Look what just a few snowflakes do to Washington, D.C."

Michael tried to laugh.

"So," Michael wrote in the back of his notebook. "The Window of Vulnerability begins to withdraw from the picture. Like a wave, bound to return, it now starts to recede, having thrown the MX missile up on the beach. New accuracy, capable of what? The precise application of the confusion of war? New York, or — ?" He looked up from the book. He had just understood what Holloway had been suggesting. He wrote "Moscow!"

The telephone rang again. Holloway went to talk. Michael opened *Strategic Review* and glanced at a review of a book about United States weapons policy in space. He was restless, and the review was thick with acronyms; they stood in his way like foreign-language roadsigns, unintelligibly forbidding. But at the last paragraph he sat up in his chair, shook his head, and bent his eyes more intently to the magazine. The author, Dr. Angelo M. Codevilla, was on the staff of the U.S. Senate Select Committee on Intelligence. In final remarks damning those who do not wish to place weapons in space, he had written:

> There is more here than mere trahison des clercs — although petty bureaucratic treason is of course rife. The arguments for using space (or indeed anything else) to fight, survive and win wars simply do not make any sense to people who throughout their careers — whether out of expedience or conviction — seemingly have traded on the premise that war can no longer be won or survived, or in any event should not be approached as a winnable, survivable proposition. It may

be that the ideological opiates of the past fifteen years have irrevocably closed the minds of a generation of policymakers, bureaucrats and top military officers. There is, however, faint hope: people are available who are either older or younger.

Michael shut the magazine.

Just before he left the interview, Holloway showed Michael the contents of a large back room. There, filling the place, was a magnificent layout of model trains.

Holloway surveyed the expanse of track and rolling stock with a look of pleasure. "What I'm getting interested in now," he said, with zest in his voice for the first time that afternoon, "is live steam."

III

From the inside of his car, with the windows closed and the engine roaring, the fighters seemed absolutely silent. They soared above a target range north of Las Vegas, tiny war machines, destroying something on the desert surface. Michael, on the way to the Nevada test site's Engineering Test Bed for the MX missile, watched them through the windshield, leaning forward to look up, his attention drawn irresistibly from the road. They were almost invisible; they were impossibly swift; they were tiny black cruciform shreds of metal thrown across whirled clouds and blue. One at a time they rose up out of a mountain, each like a flake of intelligent and vicious shrapnel, soaring up and overhead on a windblown trajectory, rising faster than a fired shell, falling faster than a bomb. In a long, implacable arch they came, one by one, to plunge into a cushion of air just above the earth, and whisk themselves away, each leaving behind itself a spark of impossible light on the surface of a shining white dry lake.

Little black crosses — up they came, soared so high, fell so fast, vanished — the lake flashed and a small white cloud drifted south. In that vast land those tiny black crosses ruled both sky and earth with their anger.

The highway curved away, and Michael followed it blindly. The battle raged on. He left it behind. He drove on northwest through a country of Joshua trees, mesquite, and bare mountains. Indian Springs. Cactus Springs. Along the road were trailer camps, a huge prison, and rows of signs parallel to the highway that built mental fences for anyone close enough to read: "WARNING, unexploded bombs." Some of the signs had bullet holes in them.

He turned off the highway at the town of Mercury, and made his way past another series of signs: Turn Out for Explosive Carriers; Busman's Wash; Danger, Power Line Xing; and A Theft from Your Government Is a Theft from You. The last bore a drawing of a man carrying off the state of Nevada. He drove for about twenty more miles on a narrow and fading road through more bare mountains until he came over a pass and down into a strange basin in which derelict buildings lay around on the sand like scattered ruins preserved over the years by dry air. Near an abandoned guard station a sign said in bleached-out colors: "Nuclear Rocket Development Station."

He stopped at the guard station and rolled down his window, in case there was someone there to check his badge. He turned off the engine of the car. Silence clapped down over him like a glass dome. The guardhouse stood at a crossroads. There was no traffic. There were no cars anywhere. The basin seemed empty. All those scattered buildings — pale, sand-washed rectangles piled upon rectangles, bristling with antennae — were deserted. Through the window the inside of the guardhouse looked dusty. The wind shook his car and sent streamers of yellow dust racing south across the basin. Michael rolled up his window and drove on.

The Air Force major at the test bed wore short sleeved fatigues in the heat of the summer day, and felt luckier than the Army men, who had long sleeves. He was a dark, slender man with a mustache and a wry, genially suspicious eye. Michael found him in the one structure on the valley floor that looked alive; a big blue block of a building with bureaucratic yellow halls. The major, whose name was William Jacobs, had been commuting the two hours from Las Vegas since the spring of 1979, when Crabtree had received his new direction and had sent the word out: Multiple Aim Points, Vertical Silos! "We had to do two things," Jacobs said, "demonstrate that you could have a transporter that could move over a vertical shelter, and demonstrate the egress of the missile's canister."

Michael and the major went out onto the basin in the major's van. They drove for eleven miles on flat roads. At last, about two miles from a little desert town called Lathrop Wells, they came to a broad area of disturbed earth and unusual equipment. There was a trencher: a large piece of machinery that seemed to be made entirely of toothed buckets. There were a few contractors' trailers. There were a couple of pieces of equipment that were so odd and clearly functional in obscure ways that they could only have been made by Ray Hansen's company, which they were. In one area lay what looked like four pieces of the buried trench: They were huge pipes, of a similar diameter, only instead of being lined with silver paper they were lined with three-eighths-inch steel. Each was flared to an extra thickness of concrete at one end. They looked like four gargantuan seventeenth-century cannon barrels lying on the desert, left there after some foundered vessel of extreme size had rotted away beneath them.

The test bed's famous machine was not hard to spot.

"So there it is," Michael said.

"That's it," said the major.

"So, that's it," Michael said. "The famous transporter."

It stood parked on the sand; a million pounds of truck resting on twenty-four tires, carrying a white cylinder, which would have held the missile. At each end was a radiator with room to accumulate several million bugs, under a cab in which one could entertain all the Joint Chiefs of Staff. The transporter had been put together with the reworked assemblies of two Terex Company earth movers, and the double-headed appearance that resulted made it look as if the purpose of the vehicle was to stretch the white cylinder.

"So that's it," Michael said, as if he had to reassure himself. "Can I get out and take a look at it?"

A few months before, a delegation of congressmen had come to see the transporter; some of them had wanted their pictures taken next to the tires. So Major Jacobs was prepared for anything. He shrugged. No objection. Michael and Jacobs got out into the wind.

The ground was utterly flat. Off in the distance to the north blue-black mountains rose out of a windblown haze. Cables crossed the flat between trailers under rows of sandbags. The rest of the strange machinery stood off to one side like retired combines in a farmer's yard. In the distance the four concrete cannons lay peacefully on the flat. Michael looked at the transporter. A ladder led up to the cab; a sign said "Notify Driver Before Climbing Ladder." Michael looked back at Jacobs.

"I fell off a dragline once," Michael said.

"Oh?"

Michael did not climb the ladder. He walked the length of the machine. At the other end, which was apparently the front, a makeshift steel framework held the wheelless cab of a pickup truck just off the ground to the left and front of the transporter's radiator.

"We're trying to figure out where the best place for the driver is," Jacobs said.

"Oh," Michael said.

The cab looked like a toy. Michael walked around the transporter again. Directly underneath it was a circular pad of concrete set into the ground; it was the plug — the door — of a vertical silo. Michael walked backwards away from the big truck.

"There it is," he said aloud. "That's it. Yup. Sure is."

He got back into the van. They drove away.

"Tell me about Bill Crabtree," Michael said.

Jacobs grinned. They were back in his yellow office.

"You been to engineering school, or something like that?" Jacobs asked.

"No."

"Oh, well." But Jacobs was moved. He had to overlook that enormous obstacle. He thumped his fist on the table. "Colonel Crabtree," he said, "typifies the real . . . ah . . . gung-ho engineer. He's so enthusiastic. He seems so dry about things. But he's very witty. He can be quite witty."

Jacobs paused. He considered Crabtree, struggled with an anecdote that was supposed to show how witty Crabtree was, but trailed off before the punch line. Then he said:

"Crabtree is a young colonel. He may not look like it, but he is a very young colonel."

Jacobs had to attend a meeting in an adjacent office. He gestured toward a shelf full of loose-leaf notebooks. "I'll be out soon," he said. "Take a look at those."

The books were full of photographs, all 8×10 color shots of operations connected with the transporter or the vertical silos. None had captions. There were shots of the Transporter-Erector being assembled in Seattle at the Boeing company; of it arriving on flatbed trucks — Jacobs had said it took three train cars to carry it. There were shots of the transporter raising the white cylinder to the vertical to lower the missile's canister into the silo with an enormous winch. There were several photos of jackrabbits. There were shots of the Transporter-Erector driving around an egg-shaped track on various surfaces. It did not do well in soft sand. There were shots of the DC-3 aircraft that Boeing chartered to fly its employees from Las Vegas to Lathrop Wells every day to go to work: a heavy old plane painted blue halfway up its side so it looked like a boat. There were photos of a shining corrugated Quonset hut and then photos of an enormous explosion, then photos of a mangled Quonset hut. That had been a test of silo strength, using conventional explosives. There were a couple of pictures of a group of eight men jogging in shorts and T-shirts beside the Transporter-Erector. They were followed by a photo of the same men lined up, two of them holding a sign which read "BMO–Shamrock–Boeing Athletic Club." Michael didn't know what the inclusion of Shamrock in the partnership meant, but guessed it had something to do with luck.

The books also contained photo after photo of the

Vertical Silo Egress Test. The test had taken place on August 23, 1979. It had been one of the more important occasions on the MX test bed. A substantial group had assembled under a windy sky to find out whether the canister would come out of its hole.

Michael studied the photos.

The summer of 1979. Hard sunshine, little tufts of cloud, a set of grandstands, a delegation of VIPs. Reporters fighting the windblown pages of their notebooks. A cameraman up in a crane. An ambulance. Gantries bearing slow-motion cameras. A flat-topped mound of earth, simulating debris, marked on the top with a white chalk grid. The mound fenced off by a festive little ribbon of orange engineer tape tied to stakes held upright by sandbags. A northwest wind. A feeling among some of the visitors that they have been transported to a different planet. A feeling of anxiety among some of the Westinghouse subcontractors who built the canister eject mechanism. A feeling among some of the Air Force officers, particularly Colonel Larry Molnar, who is up from BMO, that they had been here before.

Cameras whine, the engineer tape flaps, hot dust blows across the desert, the guests sweat. Attention becomes devoted to the center of the flat-topped mound. The cameras, the engineer tape, the chalk lines on the light brown earth, and particularly the grandstands full of shirtsleeved visitors, make it appear that some kind of odd sport is about to occur here, or some demonstration of combat.

The visitors lean forward in anticipation.

Suddenly the earth in the center of the mound heaves gently, and a huge yellow tube rises straight upward, lifting a neat plug of earth, a sort of cap, a meter thick and three in diameter. Dust squirts from the side of the

tube as if under pressure. The tube grows longer and
longer. It seems to lift itself in little jerks: It rises half
a meter, appears to bounce slightly as if gathering itself
for this great effort, then rises another half a meter. The
canister heaves itself up, then up, then up, little bits of
dust blowing off. It reaches its full height, perhaps four
meters. It stops moving. There is a little puff of dust or
smoke from a seam about halfway down the tube. Slowly
the top section of the canister, bearing the cap of earth,
tilts sideways until at last the whole upper piece falls to
the ground. The top half of the cylinder stands askew
beside its larger sibling. A short cable connects the two.
All motion ceases. The dust blows away. The cameras
wind down.

There is a bustling sound of approval from the audience, and smiles and good cheer among the officers.
Westinghouse officials shake hands with Air Force brass.
Civilians look at each other. Everyone is talking. Hey,
look at that. Can you believe how it picked up that dirt?
Fell off just as planned. Good, good. Fast, wasn't it? Hot
damn! The vertical silo is going to work!

Just before Michael left he asked Major Jacobs about
the track team.

"Yeah," Jacobs said, "we competed against the Norton
Track Club."

"What was the Shamrock?"

"Oh." Jacobs grinned. "The Shamrock is a, a — ah,
bordello — down at Lathrop Wells."

IV

Michael went back to Vandenberg, to see a missile fly.
He was told to arrive no later than 8 P.M. and to get a
good sleep. He was given a room in a little suite in the

visiting officers quarters, made in an old barracks building. It was a strange room, reached up steel stairs, which smelled of its new carpet, had a plastic walnut-grained headboard on its twin bed, and had three genuine pallet knife and brush oil paintings hung on the walls. Michael went to bed early and tried to sleep, but someone came in to the suite at 2 A.M., quietly, turned the heat up to approximately the level of a steam bath, and went to bed, leaving Michael dozing in his own sweat.

In his half sleep he dreamed of a yellow schoolbus hurtling through space while the driver threw sharpened children out the windows. At 4 A.M. he was awakened by the empty click and crackle of a digital clock radio supplied by the Air Force. He showered and shaved in a huge, cold bathroom, and went down to the base reception center to meet his escorts. The head of the group was a Major Grellman, a short, slender, terse, self-possessed man, like so many Air Force officers Michael had met. Major Grellman and Michael were soon joined on the parking lot by a group of captains and by two Swedish journalists who looked haggard. They had come to take photographs. Their car — and the jacket of one of them — bore the insignia "Reportage." Michael, Major Grellman, a captain named Davis, and the Swedes clambered into a blue Air Force van. The van zoomed along wet roads.

Captain Davis talked about his wife. She had been sitting on the beach somewhere near Vandenberg, he said. The weather had been perfect: a gentle surf, an orange sunset, bright water, a freighter going past in the radiance of the sun. "She thought that she just needed one more thing to make the scene perfect," Davis said. "That was a missile going off. Then, just as she was thinking that, a missile went off."

The van followed a police car that seemed to be in a

hurry. In the indistinct dark horizon green rotating beacons flashed: They were launch sites that were not in use. Davis pointed out a cluster of lights on one side of the truck and explained that the crew of the missile had been on alert for ten days (in shifts), waiting to send off their weapon. The missile itself had been shipped from a base in Wyoming some time before, having had its three RVs removed and replaced with identically shaped and weighed radio telemetry packages. The missile had also been armed with explosives so it could blow itself up. Few had to be blown up, Davis said, although one, in an apparent fit of pique, leaped out of its shelter, spun around angrily, and tried to go right back into the hole.

Davis was the talker of the group. He was a straightforward man, with glasses and a stolid, responsible look.

There was a beauty to this business, he said. Sometimes he thought about the beauty of nuclear war. He was once stationed with a missile squadron near Great Falls, Montana, where silos are scattered over an area where visibility is fifty to a hundred miles. For a little while there the war would be strikingly beautiful.

"If I could separate myself from what was happening," he said, "I often thought that I would like to go up on the mountains and watch. Seeing fifty missile plumes all at once would be beautiful."

Davis's regular job was teaching new Air Force missileers. He told them what atomic weapons did, showed them photographs of Hiroshima and Nagasaki, and asked them if they had any reservations about performing their duty. Those few that did were transferred.

The van went down a dirt road past low bushes and ice plant. The officers were jocular. The Swedish journalists said absolutely nothing; it was foggy.

The van arrived in a small cleared area in the brush

where about twenty-five other cars were parked. It looked like a California teenage keg party; except the muffled voices that hung in the damp air were neither aggressive nor amorous. The Swedish journalists got out of the van stiffly. Major Grellman said: "There's about five minutes." "Oh, my gosh," the photographer said. "Five minutes." "That's the launch facility over there," Major Grellman said, pointing to a faint glow in the fog in a direction that felt like north. "Aim your cameras there; you might get something."

The Swedish journalists rushed off, trailing tripods. Michael and the officers hung around the van. Several of them, looking up at the cool sky in distaste, got back in the van. Captain Davis explained to Michael what was happening in the control center about two miles away. The launch crews which had been waiting for ten days for this moment were now receiving a coded message from the United States Strategic Air Command. The message would say: "Fire!" Two members of the team would turn two keys; a control panel would be changed, then two other members would turn two other keys. The missile would depart for the sea around Kwajalein Atol, in the South Pacific.

There was a sound of surf, and in the distance came the sound of the speaker in the police car, which was parked on the road. That made it seem even more like an illicit teenage party. A few other voices came out of the fog, and once someone's large flashlight cast a short, wide, dense shaft of light upward. The air smelled of salt water and kelp.

Suddenly the fog went white. There was no sound; there was just an immense light — which came from a direction that felt like west. There was laughter from the van as this was pointed out, loudly, to Major Grellman. The light filled that quadrant of sky, but there seemed to be no source. It just lighted the world, and silhouetted what looked like about fifty people standing on a small hill, facing the light. The people did not move, although there was one voice raised in a shout that sounded halfway between triumph and pain. Like the light, the shout had no visible source.

Except for the shout, everything was silent: The light was a dazzling change brought to the earth. It was entirely white, and the people silhouetted against it seemed not to be black, but to be a hazy whitish-gray themselves, as if the light drove right through them. They stood on the ridge, immobilized by the light, and they looked like a line of worshipers; like a representation of all worshipers ever, standing in awe of the incomprehensible.

The silence lasted about five seconds. Then the sound came, not a slow building of sound, but a roar that began with a slap. It sounded as if the roar had always been there but that a shutter had opened with a click and allowed it to be heard. It was a ragged thunder; not white sound, like a waterfall, but a rough, crackling, thumping roar. It was like an enormous fall of boulders, the underlying booming highlighted by the smashing of the individual rocks.

Michael stood still beside the van. As the light rose in the mist like some soul going out of the earth itself, he shivered. There was drizzle on his neck. Sweat gathered in the small of his back. The light ascended. The noise began to fade into a precarious quiet.

The light climbed away. The roar went with it. The missile was gone. There had been light and a noise, and now the light and the noise diminished.

Quiet voices began to be heard.

"It is usually more impressive than that," Major Grellman said apologetically.

An odd sound came from the inside of the van. All the captains were laughing.

"There it goes," one of them said, "our Stealth Missile."

"Hey, tell the photographers that. Maybe they can make a joke of it. They'll need something. Came all that way for THIS. Stealth Missile."

The Swedish journalists returned from the direction of the little glow in the sky, trailing tripods. They did not have much to say.

"That was our Stealth Missile," said one of the captains. The others laughed. The Swedish journalists got back into the van. Michael, Major Grellman, and Captain Davis got back into the van. It was again night. The missile had gone. Michael, eyes huge in the darkness, looked out the window, keeping to himself. The missile had gone: To where? The sea? The desert? Durango, Spain? Or someplace marked on someone's office map with a red-headed pin — somewhere on the broad face of Russia?

As the van drove out on the shiny roads back towards the base, Captain Davis said:

"It'll be there in about twenty minutes."

23

I

TIME: 9:05 P.M. Eastern Daylight Time.

September 1979. Hurricane David. A thousand people die in the Carribean. Wind and waves do sixty million dollars' property damage to Florida. The storm smashes up the East Coast: Virginia, Maryland, Delaware, New Jersey. Land disappears under abnormal tides. The moon goes red in total eclipse. Land is washed away by the sea. People die in storm sewers. Rain floods Washington. Branches of trees blow through the streets. Marvin Atkins drives in from Baltimore on the Tuesday evening after Labor Day. The Interstate is a wild, dark place of blowing mist and rain. He drives into the parking garage of his apartment with relief. He gets out of the car. He looks around at unexpected darkness. Most of the power is off in the building. Atkins's apartment is on the thirteenth floor. He walks up the stairs. He reaches his door at last, gasping. He unlocks the door, feeling the sweat cool on the back of his neck, like rain. Inside, the telephone is ringing.

"Hello."

"Marv!"

"Hi, Sy."

"Where the hell are you? We need you. The President's announcement."

Time: 9:12 P.M. Eastern Daylight Time.

II

Time: 9:51 A.M. Eastern Daylight Time.
President Jimmy Carter:

I have a statement to make about the new strategic deterrence system which I consider to be quite significant. Some analysts would equate it with two other major decisions made by Presidents in this century: The first, to establish the Strategic Air Command itself under President Truman, and the subsequent decision by President Kennedy to establish the silo-based Minuteman missile system.

... I decided earlier this year to proceed with full-scale development and deployment of a new, large mobile ICBM, known as the MX. I made this decision to assure our country a secure strategic deterrent now and in the future. . . .

[After] full consultation with Secretary of Defense Harold Brown and my other principal advisers, I have decided upon the following configuration for basing the MX missile system. The MX will be based in a sheltered, road-mobile system to be constructed in our Western deserts. . . . This system will consist of 200 missile transporters or launchers, each capable of rapid movement on a special roadway connecting approximately 23 horizontal shelters. . . .

[As] President, I have no higher duty than to ensure that the security of the United States will be protected beyond doubt. As long as the threat of war persists, we

will do what we must to deter that threat to our Nation's
security.

Time: 10:02 A.M. Eastern Daylight Time.

III

Time: 12 noon, Mountain Daylight Time.

Michael sits on the floor in a friend's living room in
Montana and watches a shadow of President Ronald
Reagan in a television that barely picks up a distant sta-
tion.

> We will not deploy 200 missiles and 4,600 holes, nor
> will we deploy 100 missiles in 1,000 holes. . . .
> Instead, we will complete the MX missile — which is
> much more powerful and accurate than our current
> Minuteman missiles — and we will deploy a limited
> number of the MX missiles in existing silos, as soon as
> possible. At the same time, we will pursue three promis-
> ing long-term options for basing the MX missile and
> choose among them by 1984. . . .
> It's my hope that this program will prevent our ad-
> versaries from making the mistake others have made
> and deeply regretted in the past: The mistake of under-
> estimating the resolve and the will of the American
> people to keep their freedom and protect their home-
> land and their allies.

Time: 12:14 P.M. Mountain Daylight Time.

IV

Time: 10:23 A.M. Eastern Standard Time.

Marvin Atkins's office, the Pentagon. In the anteroom
one of the secretaries is typing fast, leaning forward in

irritation to make corrections, typing again. T. K. Jones, a former Boeing executive who has taken Seymour Zeiberg's place, comes through the door. He is tall, with wavy gray hair. He says: "Is it ready yet?"

"No, Mr. Jones."

Inside the office Michael and Atkins sit at the end of the long table. The dart board is still on the wall. The signs on the files say CLOSED. On top of the files are what looks like a pile of model RVs. Atkins laughs:

"One could have watched the key indicators and drawn correct inferences. For a month before the final announcement, Senators Paul Laxalt and Jake Garn made very negative statements about any kind of MPS. I interpreted that to mean that the President had promised them that there would not be an MPS in Nevada and Utah. They were very loyal to Reagan."

Atkins laughs. Michael laughs. Atkins is playing with a paper clip. Michael is scribbling vigorously in his notebook. He stops, and looks up.

"This all depends on friendship, personal integrity, trust, doesn't it?" Michael says. "Us. . . . We civilians can either react to you with cynicism and say you're all a bunch of crazies." He laughs. Atkins laughs, then cuts it off. Michael continues. "Or we can react with trust, right?" Atkins watches him coolly. Michael says. "So how do you develop that sense of trust?"

There is silence.

Michael says: "Do you have any suggestions?"

Michael laughs. Atkins laughs. Color leaps in his cheeks. Atkins answers:

"All of us would agree that there are a lot of problems in this business — mismanagement, incompetence, and so forth. And that's precisely why a lot of people have decided to stay on the government side. Because they thought the overall job was important, and they thought they could help. To some extent that's ego satisfaction, but I would say the overall situation has improved tremendously."

Michael looks at Atkins. Neither laughs. Atkins says nothing more. Michael flips a page in his notebook noisily.

"What's going to happen now?"

They laugh again.

"It's hard to wrap it up when the story never seems to end," Atkins says. He laughs. "I was watching television recently, and there was a program on natives on Bali. They explained that the people on Bali have a different concept of time from ours. Where we view time as proceeding in a straight line, they say that everything has cycles. A cloud has a short cycle, a rock has a very long cycle. Everything moves in cycles. I said, 'Gee, I understand that. I fit right in.'"

Atkins laughs. Michael laughs. Atkins eyes are cool.

Michael walks rapidly through the halls of the Pentagon. He passes a wooden table outside a door. A sign is taped to the table: "This is NOT surplus."

Time: 10:45 A.M. Eastern Standard Time.

V

Time: 4:30 P.M. Eastern Standard Time.

Michael is waiting in the anteroom of Jasper Welch's office: Research, Development and Acquisition. The Pentagon, fourth floor. Welch, the physicist in uniform, the man with the drowsy, alert eyes. A narrow box full of earth and green plants is at his left elbow. Michael studies the plants. Michael says to the secretary:

"There's a pencil growing in your plants."

She looks up.

"Oh, yes. That's where we grow all our pencils. Would you like me to grow something for you?"

"Well, I'm always losing my ball-point pens."

A lieutenant colonel who occupies the desk over by the window says:

"We can do anything. We're R and D."

There's a pleasant silence. The lieutenant colonel says:

"I hope Ivan has that on his tape."

Welch comes out of his office and lets Michael in. He walks as if all his muscles ache. He looks sleepy. His eyes watch Michael. As they talk his voice, which started at a normal volume, becomes more and more quiet. They talk about the Reagan decision, about the Townes Committee, about Albert Latter, about Bill Crabtree, about the effect of military mafias on policy and careers. He traces each of the names Michael gives him, all friends, back to their professional groups of origin. Latter — the Second World War science mafia; Crabtree — the Guidance mafia; Toomay, the Radar mafia. He speaks fondly again of the Kirtland mafia. He mentions what seems to be a small part of Jimmy Carter's decision back in 1979 to go ahead with the MX.

Just before the big moment there had been an in-house debate over the size of the missile. Bill Perry, the Gentleman of Science, wanted a smaller weapon than the 92-inch-diameter Air Force design. He preferred an 83-inch MX, "not because it was less ostentatious," he wrote later, "but because it could be used for both Air Force and Navy programs." People like John Toomay, who would retire on September 1, a couple of days before the final decision was announced, also preferred the smaller missile for two reasons: It would be easier to hide and it would be more clearly a weapon of deterrence rather than aggression. The 92-inch missile was the weapon requested by Bruce Holloway, which would allow the most precise and devastating application of the chaos of war.

But the issue was not settled on those lines.

"Zbigniew Brzezinski (Carter's national security advisor), I think, gets the prize for settling the missile size," Welch says quietly. "It was a very nip and tuck thing. The Air Force engineers had grave technical reservations about the notion of a joint development with the Navy. It eventually came down to a decision because Brzezinski reasoned that it would not do politically for the U.S. to build a missile smaller than that allowed by the SALT II treaty. Carter's image would suffer."

The long room is quiet. Welch's eyes are tired. It is late in the afternoon. Michael absentmindedly draws a large, upright outline of a missile on the page of his notebook. Noticing what he has done, he abruptly turns the page. He changes the subject.

"Do you have any way of describing all these people we have talked about?" Michael asks. "A common characteristic which you all —"

Welch breaks in abruptly. His voice is suddenly almost loud.

"I would say that the common characteristic that a lot of people miss is just a whale of a lot of super patriotism."

"None of you seem to talk about that much."

"That's right," he says, thoughtfully. His voice again is soft, so soft that Michael, later, will barely be able to hear it on the tape. "Because everybody does it."

Welch has visitors waiting. Michael rises to leave. At the door he suddenly turns to shake hands with Welch, and surprises him. Welch's right hand is full of lighters and cigarette packages. He reaches out for a moment and grips Michael's right hand with his left. From the hall it must look like a gesture of reassurance. Perhaps it is.

Time: 5:10 P.M. Eastern Standard Time.

VI

Time: 10:35 A.M. Eastern Standard Time.

Sunshine on the peaceful town of Punta Gorda, Florida. A breeze on the inlet. Pelicans over the water. John Hepfer sits in his living room gently punching a little pillow on the couch.

"I wasn't too surprised by the decision," he says. "I saw on one of the evening news shows Paul Laxalt coming out of the White House. The reporters were there and they asked him something about the MX, and he said, 'The President's not going to buy it.' Since Laxalt happened to be a very close friend of the President, I figured when you're in that job in the White House you have to lean on somebody fairly closely. Your friends are going to be fairly influential."

Hepfer puts the pillow down. It is very quiet. Hepfer has already shown Michael the Heathkit microcomputer he built when he retired. Now he talks about the crazy months after Carter's announcement, when the Nevada-Utah basing system exploded in unexpected controversy.

"You know," he says, putting his fingers together and smiling briefly, "at the time we first looked at Nevada there were only nine hundred and some thousand people total in the state, total. We had had classified projects out there in the desert for years and nobody even knew it. We flew over that area and up and down the valleys, and we couldn't see any people."

He talks about the evolution of the design of the machine that would carry the MX on its back all the time, parking in shelters and driving out to launch, to a truck that would unload the missile at the appropriate garage. Perry and Brown had favored the first device, which was even larger than the transporter the Air Force had already built, which also required an even larger device, a hangar on wheels, to hide it as it moved. "Ol' Crabtree kept needling them down, driving them down, down, until we felt we had something acceptable."

The switch was announced unexpectedly at a hearing by Harold Brown, who at the same time told the senators that the oval racetrack deployment had been replaced by a straight-line design. The road change won a banner headline in the Washington *Post* and a new nickname, Dragstrip. Nobody had noticed the change in vehicles. (Welch: "We were delighted.")

Hepfer exercises his grin sparingly. He and Michael laugh about Michael's observation that four years after the trench was killed a wire service is still describing the MX system as an underground railroad.

Hepfer laughs gently.

"I got so I couldn't stand to read the newspaper, they were so inaccurate. You know, you get in those jobs, and you get pretty wrapped up in your own little world. You're all pretty dedicated, as you have probably heard. You don't realize it, but you sacrifice a lot. Your families, things in your personal life, things you like to do. I was sitting there one Saturday morning, and I thought, Well, I could keep doing this. I guess I sensed that they were going to recycle the whole thing again. I wrote the chief a letter and told him I'd like to retire."

He picks up the pillow again and kneads it. He looks out the windows at all those other homes scattered along the inlet, standing silent in the breeze.

"It was kind of a — hasty thing. I shouldn't have done that without going in and talking to him. They had all been very good to me. They'd let me do most anything I wanted to do. Nobody'd bother me."

Hepfer smiles. His big grin stretches clear out to his big ears. "It was like I had my own little Air Force."

Time: 10:55 A.M. Eastern Standard Time.

VII

Time: 1 P.M. Eastern Standard Time.

Inside the lobby of the Martin Marietta main office in Orlando, Florida, there is a full-scale mockup of the Patriot missile, an advanced surface-to-air missile being built by Raytheon Corporation and Martin Marietta. The Patriot is about the length of an outboard motorboat, but slender. There are dents and scratches in the mockup.

At a semicircular desk sits a man with blue-gray hair and lively blue eyes, writing quietly.

Michael goes up to the desk to sign in. "Hello!" the man says cheerfully. "Who do you want to see?"

"Dr. Seymour Zeiberg."

The man looks up at him with his bright, happy eyes.

"Dr. Zeiberg," he says. "What a name!"

Michael says, "Um."

The blue-eyed man writes "Zeiberg" on a check-in sheet.

"Dr. Zeiberg!" he says. "Now *that's* a real *scientist's* name."

"Um."

The man gives Michael a pass on a clip: "Visitor. Escort Required." Then the man looks fondly at the check-in sheet as if he has just created an unusual piece of artwork between the ruled lines. His eyes glow with delight.

"Dr. Zeiberg!" he says. "Now that's the kind of name you find in a *science fiction* novel."

Zeiberg's office is empty. There is a small desk and a large round table. Michael sits down at the table. On the walls are a couple of arrangements of metalwork mounted on dark boards: one is of welded foliage, the other of old tools: hammer, wrench, meat hook. On the desk is a large black satchel with worn brown seams, a few papers, a container of sharpened pencils, and a familiar hand grenade, the pin installed. In the center of the round table is the little gong.

Michael waits. The room looks out on the construction project, a parking lot, and a wonderful brightness of Florida sunshine. Michael examines the legendary gong. It is about the size of a child's fist. It is made of dull brass. It hangs from a brass holder. A small drumstick rests beside it. Michael looks around. There is a murmuring of voices outside the open door. He studies the gong. It bears parallel decorative scratches. It is slightly convex.

There is no sign of Zeiberg. Everyone outside is preoccupied with other things. No one is walking past the window.

Michael leans over and strikes the gong firmly with his fingernail. It makes a very small clank. It has about the same resonance as an aluminum spoon. Michael sits back. The gong swings on its stand for a moment and then is still. Michael grins. His eyes alight, he looks over at the grenade.

Zeiberg comes in.

He looks different — younger, happier, warmer? He is working in Florida by himself while his family finishes the school year near Washington. He is now vice president for engineering in the missile systems division, the boss of about twelve hundred engineers, making things like the Patriot missile. Zeiberg's face in the color photo on his badge looks radiant.

"How is your work going?" Zeiberg says, settling comfortably on the other side of the table.

"Oh!" Michael says. "I don't know. It's — it's progressing."

"You're almost at a point where you could wait it out

and publish just before the —" Zeiberg laughs "— the ostensible big announcement, which won't happen, I'm sure."

"Which won't happen?"

"They won't have enough work done to make it happen."

"By July 'eighty-three? You don't think it will?"

"No. The options they're looking at are not yet under contract. Things are dragging. Congress refuses to let them work on the continuous patrol aircraft, and they're bickering over that." Zeiberg seems delighted. "It's pretty chaotic." He grins widely and chuckles.

"A lot of the people who played with the system in the political context, highly respectable members of the defense community, people at the Hill who played with the system, poking fun at it because it became a surrogate for Jimmy Carter, poking fun at Jimmy Carter's naiveté on SALT, suddenly realized they had brought the house down." The laughter has gone out of Zeiberg's face now. It has become bland, unemotional, serene. He goes on, steadily.

"Then, when less knowledgeable people then took charge of the Defense Department, they looked at it and said, Oh, these guys are poking fun at it, it clearly is a dumb Democratic system; we need a nice Republican system. That's a critical essense here."

Zeiberg's eyes are steady behind the black rims of his glasses. His left hand is at rest on the dark tabletop. The hand, Michael notices incongruously, was very clean.

"The Townes Committee," Zeiberg continues placidly, "was pretty bad."

"Hm," Michael says.

"They made a mistake of developing what I call the green grass syndrome," Zeiberg says. He picks his left hand up from the table and folds it with the other across his stomach. "That is, there must be a better way to do the job. While they looked at all the work that had been done — they had voluminous briefings — they acted as if there was still something better out there."

Zeiberg leans comfortably back in his chair. So does Michael.

"The decision was clearly and entirely a political judgment. It was unfortunate that President Reagan got some pretty bad advice."

Michael finally asks: "What was your personal reaction. Did you . . . ?"

"Oh, I was pretty disappointed," Zeiberg says easily. "I'm not entirely — I haven't given up entirely." He smiles. "The way these things go, and the way this team seems to be pfumfering. I have some feeling we can reverse that decision, to some degree." He grins again, that huge smile that, frozen on his badge, looks so entirely full of joy.

"And I still work around the shadows in Washington when I can."

Michael laughs. "OK," he says. Zeiberg chuckles, then adds:

"I'm still — a — member of the MX mafia."

Time: 2:08 P.M. Eastern Standard Time.

VIII

Time: 5:10 P.M. Pacific Time.

Wind chimes. Low clouds off the Pacific. Big John Toomay's front yard is all dug up. His wife is replanting. There are golf clubs by the door. Toomay, all two meters of him, sits sprawled in his couch.

"Well, what have we decided about these guys?" he

says. "We think that Hepfer terminated his career early because of the White House's whimsical redesigns — its ruthless exploitation of all the talented people that he had. We think that Crabtree's career has been helped regardless of the fact that the MX is at least a semi-failure. Because there's an essential philosophy in the code of the military guy. I do my job to the best of my ability; it's a job determined by my civilian betters, and if things don't come out right, I've done all I need to do. Crabtree will have a very healthy cynicism about our civilian betters.

"OK. We agreed that Zeiberg, subconsciously or consciously, wanted the MX as his monument, worked incredible hours to achieve that, made a number of compromises, perhaps even with his own technical principles, and then didn't survive the political thing, and the program didn't survive either. He may be more disappointed than anybody, because he tried so hard and it didn't work. On the other hand, we engineers have something like the philosophy of the military guys. We don't blame ourselves. So I don't think he'll be psychically affected. What I'm saying is, everybody's going to rise above this whole thing."

Toomay sips on lemonade. Michael sits relaxed across the low table. There is a book on the table: *The Dancing Wu Li Masters*. For a few minutes they talk about the proposal for MX called Dense Pack, a complete turnaround from scattering the missiles across vast areas to jamming them together in a bunch so close together that incoming warheads will kill more of their brethren than U.S. missiles. Another invention! "The catch," Toomay says with a trace of a smile, "will be getting Dense Pack past the Community of the Nuclear Physicists, the Electromagnetic Pulse Community, the Hardess Community, the Tacticians." All those mafias.

"But Dense Pack," Toomay says, "may not be dumb."

There is a pause. Michael leans forward, his elbows on his unused notebook, his face calm.

"We hear so much . . . so many people selling the threat," he says, "companies like Rockwell, everybody, telling us what to fear. Who are we to tr——" Michael paused. He says it another way. "Who are we to believe?" He puts his chin in his hands and watches this uncommon man ramble on.

"Oh, yeah. I join those that view that with extreme cynicism. Because on the face of it these companies have a conflict of interest. You know, North American Rockwell is famous for that, and how can anyone believe what they say when they're going to make billions on the B-1? Can't believe it. You need to find a guy who hasn't got any money to make. Well, no military guy's going to make a cent on it. But then the anti-military people formulate this other quality in the military guy, which is, no, he isn't going to make any money, but he's a man obsessed with the desire to kill us all."

Toomay sips again. He is arguing with himself. "Well, in some cases that might be true. But at least it's more useful to listen to that guy, because he's forever isolated from —"

A pause.

"— He's not totally isolated but he can't —"

Another pause.

"— Yeah he can."

A third pause.

"—I was going to say he can't jump out of his military suit into a contractor's suit, but he can. So."

Again, silence.

"So you can't listen to anybody."

After a moment, Michael sits up. He repeats a bit of news he received earlier:

"And now you're going to work for Texas Instruments!"

Toomay says: "Yeah." They both laugh.

Virginia Toomay comes in to discuss supper. Trays by the television for Monday night football. The two of them get into a cheerful conversation about the characters of military men of science versus fighter pilots.

"In the science and engineering communities," Toomay says, "they don't have the real fighter-pilot type of behavior. That doesn't mean that scientists and engineers aren't having as much fun as pilots, it just means they're not being obvious about it."

Mrs. Toomay breaks in.

"I don't think they are."

"They aren't having as much fun?"

"No way!"

"What's your definition of fun?"

"Well . . ."

"OK."

"That's true."

She leaves for the kitchen. Toomay gazes thoughtfully across the room.

Michael says:

"When I was a kid I remember hiding under a desk during an atomic bomb air raid practice. My generation has a lot of memories like that. We have a rooted fear of nuclear war." He takes a breath. "Do people like you and these guys who are responsible for preventing it . . . do they share with us . . . these vivid fears?"

Toomay considers the question. Michael hears wind chimes. He hears the wind. He hears his tape recorder running in the still room.

Toomay says:

"Yeah. There is some tendency to become callous. A lot of people are more callous than they ought to be.

But I know the guys on the highest levels are always thinking of that. They have the same images that you have, and they have the same resolve: They want to stay out of nuclear war. But there's the one condition, which is Eisenhower's condition, that there are some things that are heavier than a soldier's pack. Namely loss of freedom. They believe that some things are worse than all those casualties. Namely, to be enslaved. There's no difference to them whether they're incinerated with napalm, vaporized with a nuclear weapon, or stuck with a bayonet. Some things are worth your life."

Michael is silent. Toomay looks at his watch.

"It's six-o-four," he says.

They go into another room. The television set is on. Virginia Toomay, who has made a point of inviting Michael to dinner in advance, had cooked a rib roast. It is magnificent. He and Toomay sit in the large room with the television and watch the game. Michael jabbers along for a few moments about football, then abruptly becomes silent. He has his notebook with him. It has never left his side. He looks down at the last scrawled remark. Some things are worth dying for. He looks over at Toomay. The television set rings with cheers. Toomay is slumped in the chair. A commercial comes on. With a remote control Toomay switches the channel. There is a Kung Fu movie: Bruce Lee throwing people to the floor. Uh! Aiee! Toomay watches it with faint interest. Michael looks down at his notebook. He writes on the next line: "WAR: Some things are worth killing for."

On into the evening the room is filled with the smell of roast beef, with the sounds of people cheering, and with the unintelligible noises of violence made by the movie warrior. Michael stares at the TV.

Time: 6:55 P.M. Pacific Time.

IX

Time: 11:15 A.M. Pacific Time.

The Nevada test site. North wind. A cluster of buildings; two gantries with cameras; a two-hundred-ton-capacity crane. The wind rushes through the greasewood.

Crowds. Cheerfulness. It is a test of the MX missile cold launch system. Today a full-size replica of the MX will be thrown from its launch canister. It is another step.

A crew of men with white hard hats clusters in the door of a large metal building next to the canister, then files away, white hats bobbing along a road toward a safe area. On another side of the structures, a small group of newspeople stands on an old road, joking in the wind and sun. Michael is among them. On their way here, up the road, they all had to pass a sign that read "MAD AREA CLOSED."

The MX canister stands between supporting structures at a slight angle off the vertical, leaning toward the newsmen. It is made of graphite epoxy. It is black. It is made of three parts. The places where it is joined are white. Michael stands talking with Neil Buttimer's successor at BMO, Captain Patrick Mullaney. Captain Mullaney is cheerful, enthusiastic. "Inside the canister is a — we don't like the word *dummy* — a simulated missile, made of concrete and steel. One hundred and ninety-five thousand pounds!"

11:20 A.M.

A siren wails from somewhere inside the structure. A loudspeaker in the distance: "T minus ten minutes." Everyone has to strain to hear. Major Bill Jacobs, the same Bill Jacobs who runs the MX test bed, smiles his sardonic smile and says: "They insisted on the siren." Michael says: "It's necessary for the atmosphere." Mullaney tells the television cameramen to start filming at one minute. "After that you never know." The television newsman who has been lying on the hood of his car sunning himself eases to his feet. A cameraman in a large, black, furry beard says: "Well, it can't be any more boring than that cork thing."

Michael asks him: "Cork thing?"

"When that thing stuck up out of the ground."

"Oh. The vertical silo egress test."

"Whatever. It just came up and fell over and that was it. Boring."

At a viewpoint on the other side of the gantry, someone at a microphone is speaking to a crowd of white-hatted men. What he says is unintelligible.

The siren sounds again. Michael strains to listen. "T minus five minutes." Two jets chase each other far overhead, their contrails mingling. The wind blows through the mesquite bushes with the sound of rushing water. The faint sound of steel humming comes from the structure. A helicopter carrying Air Force cameramen clatters along in the background, standing almost still in the wind.

Captain Mullaney says with a laugh: "When it happens there will be a cloud of steam and unhealthy vapors, so if it blows this way, book it."

Michael has seen film of previous tests of the canister. In the films the dummy missile leaps from the canister, followed by that cloud. It seems to fling itself into the air, accompanied by a flock of small black objects. The objects look like crows, suddenly released from the black tube to fly with the great white cylinder. The missile reaches several hundred feet and then belly flops with a thump to the ground. The crows follow it down. The crows are pads blown out of the tube by the launch. Michael has watched it several times in slow motion.

"What is the cloud that comes out, Bill?" Captain Mullaney asks Jacobs. Jacobs says, with a trace of a smile:

"Hydrochloric acid and steam. It's a combustion product."

Michael says to Jacobs, recalling photos of the vertical egress test: "You don't have any grandstands this time."

"Nope. We are getting *austere*."

The distant speaker can just barely be heard over the wind.

"T minus two minutes."

"I watched 'em drop that thing out of the airplane back in 'seventy-four," says the bearded cameraman. Michael looks at him again. It doesn't look as if that one interested him much, either. Michael looks back at the canister.

All those years, since the air drop. Little John Hepfer was a young general. Bill Crabtree was still with the A-10. Big John Toomay was working in the Pentagon. Lloyd Wilson was still living. The Trench had not yet been born. Nor the Race Track, nor the Zipperditch, nor the Dragstrip, nor the Transporter-Erector-Launcher, nor Big Bird, nor Dense Pack.

"I was in a chase plane over the desert," says the cameraman, "when they dropped that sled thing out that was supposed to be the missile."

Michael says: "Say. I want to ask you about that. Did the parachutes come off? Did the thing fall into the desert, zip! Ka-BOOM?"

The cameraman looks at him a moment.

"Um, yeah," he says.

The siren blows again. The faint loudspeaker says, "T minus one minute."

There is a yellow light above a small building downrange from the canister. The black canister looks like the barrel of a howitzer. There seems to be a tiny strobe light flickering at the top of it. Michael stares at it. He is suddenly gripped. In a moment the missile will fly out, will blast from the desert and hang free for a moment in the wind. The MX. It's going to work. Hot damn!

He scrawls rapid notes: Chopper. Siren. Red flashing light. Sound of cameras. T minus 10 seconds.

Five, four — the wind blows the voice from the little hut away. There is no sound on the desert but the helicopter and the wind. The wind blows for five seconds; ten seconds. Someone says: "Oh, oh."

The MX canister stands, silent, pointing at the sky. The distant men in the white hats are motionless, watching. Mullaney and Jacobs and the newspeople do not move, watching. For several moments the whole apparatus seems out of control. No one knows what it will do now.

Major Jacobs lights a cigarette.

Captain Mullaney says:

"We're looking for a volunteer with a big lighter."

The radio in Jacobs car says: "It's an abort."

Major Jacobs goes over to Mullaney's rented car and sits down sideways in the driver's seat. He very gently hits his knee with his fist.

The MX canister stands, silent, pointing at the sky.

Time: 11:34 A.M. Pacific Time.

X

Time: 10:30 A.M. Pacific Time.

Michael had never seen Bill Crabtree walking. Crabtree had always been installed, like a flexible fixture, behind his desk. Now, suddenly, Crabtree himself appears in the lobby of the Air Force Space Division 110 to escort him back to his office. Crabtree is a different man on his feet. He is small, broad-shouldered, and he walks with a kind of lunge, his head lowered. He looks as if he would be willing to try to walk right through a wall.

Crabtree is now the project director for the Air Force military satellite project. He shows Michael into a tiny office almost filled with two desks set in a T arrangement. He sits down. He says: "I got your letter. I've tried to think about it."

Michael had written him after the last time they had spoken: "What I still would like to know is how do you develop that sense of service that you describe. . . ."

Crabtree sits at the desk and puts his feet up on the extended metal typewriter stand. As he talks he twitches his head and, occasionally, his feet, in emphasis. The shoes squeak. He says:

"A couple of times — I thought about leaving the Air Force. One time I even had my resignation in — I thought I might go back to graduate school — that was in nineteen seventy-one. To the extent I can remember I stayed in — because of a personal issue of commitment. It is not easy to work for the government. It is a frustrating process." His shoes squeak. "I figured if I quit, then I was quitting, and that was something I couldn't cope with."

He glances at Michael. He struggles with what he is trying to say. Michael leans forward and watches him with growing intensity. Crabtree fights with the unfamiliar self-examination. His shoulder twitches. Michael watches, his face suddenly alive with affection for this fierce man, who conceals himself so well.

"As I said — I first started looking for something that was relevant. I stayed with it — I guess — for two reasons — one just because I didn't want to quit — and the other one is that — when you're in this side of the business — you're trying to solve a problem and that problem deals with the threat — and so you get to know the threat very well."

His shoes squeak, his head twitches, he puts one arm on top of his head. His words come out in bursts of monotone:

"I guess — this may sound a little corny — but that gets — to be — fairly real."

There is a long silence. For several moments Crabtree is still. His mouth works slightly as if he is chewing his words. His right hand plays absently with a corner of a desktop drawing pad that is covered with notes. His face is smooth, and looks young, except for the trace of silver in his hair. His hand seems weathered.

"There are emotional motivations — but the thing you can point to as the factual motivation is the detailed understanding of the threat — and — I don't know the public could ever —"

Another long pause. Crabtree shrugs his shoulders and works his mouth. Then the passionate monotone continues.

"So much of the data is unavailable — to the public — and what the public does have access to is — just downright — misleading. Both ways. Not just in the understatement. Overstatement as well."

Michael says:

"It appears to me that depending on your — emotional orientation you can pretty much believe what you want."

Crabtree's shoes squeak. He says:

"Yes. It's — it's — these issues are very important and it's easy to see how someone dealing with — the data which is available particularly in the public arena, could reach just about any — conclusion of what they want to reach."

Michael leans forward, watching Crabtree, chewing on the top of his pen, his eyes bright. He says:

"One wishes there was some way you could say these are the statistics, these are the graphs, and this is the situation. But it doesn't seem to be that way." He watches Crabtree. Crabtree's shoulder twitches. Michael goes on.

"It seems that so much depends upon one's . . . perception of the people who are running the show." He smiles, his own shield down. "Which is a difficult spot to be in."

Crabtree stares out at the parking lot.

"Right. But these kinds of decisions have to be decisions by men."

Michael rushes on, his voice deliberately casual.

"So, you're put in a position where you either choose to be cynical about the people or . . . trust 'em."

Crabtree shrugs. His shoes squeak. He moves a shoulder.

"And neither one of those — is right."

Time: On the nuclear war clock drawn by the *Bulletin of Atomic Scientists*: Three minutes before midnight.

24

It was late in the day. Dr. Albert Latter closed the door to the hall and sat down in a chair with a cup of coffee. He leaned the chair back against the blackboard and sipped the coffee. All alone on the board was drawn a circle with a bent arrow through it.

"Can't see the mountains," Michael remarked, looking out into the bright gray haze. A storm was coming to California.

"Good," Latter said. "Maybe they'll get some snow."

Latter's coffee cup had a sailboat on it. He sipped peacefully. Michael sat down next to a metal sculpture of a skier made of nuts and bolts and other hardware.

"You were at the University of California during the war," Michael said. "Were you associated with the Manhattan Project?"

"Yes," Latter said. "When the war broke out I was scheduled to go to the University of California to study theoretical physics under Dr. Oppenheimer. When the war came I wrote the Navy. I was going to join and do something like radar. While I was waiting I went up to the University of California and worked with him, and I never heard from the Navy. After a while I found out that I had been impressed into this project. I didn't have a particularly significant task, but I was working with a lot of Nobel Prize winners."

"Did you go to New Mexico?"

Latter got up, carefully protecting the coffee. He stared out the west window at the invisible Hollywood hills. He told Michael what was true.

"No," he said, with a faint smile. "I heard that Los Alamos was not a . . . great place. Barren. No girls." He smiled. "So I talked to E. O. Lawrence, and I stayed at the University of California for the rest of the war." He walked across the room. "I got out to Los Alamos later, of course."

Michael glanced down at his notebook. So Latter had missed it, missed the great and terrible moment. Missed being at the heart of the greatest invention.

"Do you regret not being there?"

Latter smiled.

"No," he said. There was a pause. Then he said:

"I don't regret anything."

Michael wrote it down. There was silence in the room. Latter looked at Michael, who asked:

"Would you say that the difference . . . that from the time of the learning how to control atomic fission and fusion as the great invention — there has been a change in the way people have looked at the weapon?"

Latter looked at him with amusement. He said:

"I'm not sure I understand your question."

Michael grinned. He tried again.

"Since the great triumph of making the atom bomb," he said, "have the perspectives of people in this business changed?"

Latter got up and, carrying his coffee cup gently,

crossed the room, and searched the invisible Hollywood hills. "Of course," he said, "Some people have since spent their lives trying to put the genie back in the bottle."

He paced back across the room, eyed the closed door into the hall and stared out at the invisible Pacific.

"You know," he said, "even atom bombs get to be old hat after a while." Michael was scribbling. When he looked up Latter had put the coffee cup down and was standing there, hands jammed in his pockets, looking at him thoughtfully.

"For some people," he said, "there is not the same sense of life or death anymore."

Michael wrote it down. He looked at Latter. Latter was staring out the window. Not a matter of life and death! Latter looked for a long time. The nuclear physicist. The inventor. No longer, with his brethren, the wizards of the earth. Crabtree had said: Trust or cynicism; that's not it. "It's more complex than that. These are just people doing what they think is right at the moment."

Latter stood there, slim as a mast, his face smoothed by a gentle humor, his eyes searching. Michael watched him. They had all done what they thought was right. But the drive for the technological solution was tired. It was used up. It was no longer — relevant. Build the bomb! Get rid of it! Both solutions of mechanism; but that was no longer the matter of life and death. There was something else now to find. The matter of life and death hung on a different thread. It's in the *people*.

Michael looked at the blackboard, at the circle and arrow, which seemed to be a representation of the male symbol turned upon itself. He saw the MX missile standing in the desert, erect, symbol of all war, built by men who no longer needed combat to test their courage or their virility or to win their safety. The one characteristic that all these men shared was not the sad fact that each seemed obliged to utter the tough old code words of war, but that against the truth of their intricate lives, those words were out of character: They were false.

Latter put the cup down and stared out the window, his small eyes searching in the maze. Searching for what? For that instant of ultimate achievement, the one he had missed? Or for the one still to come, the next great invention? Michael sat up in the chair, and stared at Latter.

"Was that the high point — the invention of the bomb — in your profession?"

Latter turned back, his eyes amused.

"Do you mean from a technical point of view," he said, "or from a social point of view? Technically, perhaps it was."

"What's going to be the next?" Michael asked. "The next great invention?"

Latter stood in the middle of the room, his hands jammed into his pockets, leaning slightly back, his eyes narrowed in the light.

"I don't think there will be as important an invention in nuclear physics," he said. "The next great invention may be in biology."

"Or in the social sciences?" Michael asked.

"Or in the social sciences."

Or from some combined effort, some ultimate Tiger Team, formed in desperation at this final moment, under the deepening shadow of the only real threat. Napalm, nuclear weapons, bayonets. Gentlemen, war!

Albert Latter looked out the window. The ocean was invisible past the boats and apartments on the water's edge. Latter strode over to the door and looked out, then looked once again at the sky. White hair, hungry eyes. For what did he search? Michael, too, stared out into the huge gray sky. For what did they search? For what did they all search? For the next great invention! The unattained triumph! The invention no less miraculous, no less impossible, than the invention of the bomb and all its conveyances! To the next great invention, the invention of peace.